LONDON MATHEMATICAL SOCIETY LECTURE NOTE SERIES

Managing Editor: Professor J.W.S. Cassels, Department of Pure Mathematics and Mathematical Statistics, University of Cambridge, 16 Mill Lane, Cambridge CB2 1SB, England

The books in the series listed below are available from booksellers, or, in case of difficulty, from Cambridge University Press.

London Mathematical Society Lecture Note Series. 131

Algebraic, Extremal and Metric Combinatorics, 1986

Edited by
M-M. Deza & P. Frankl, CNRS, Paris
and
I.G Rosenberg, University of Montreal

CAMBRIDGE UNIVERSITY PRESS

Cambridge

New York   New Rochelle   Melbourne   Sydney

Published by the Press Syndicate of the University of Cambridge
The Pitt Building, Trumpington Street, Cambridge CB2 1RP
32 East 57th Street, New York, NY 10022, USA
10, Stamford Road, Oakleigh, Melbourne 3166, Australia

© Cambridge University Press 1988

First published 1988

Library of Congress cataloging in publication data available

British Library cataloguing in publication data

Algebraic extremal and metric combinatorics, 1986
1. Combinatorial analysis
I. Deza, M-M. (Michel-Marie), *1939 -*
II. Frankl, p. (Peter), *1953 -*, II. Rosenberg, I.G. (Ivo G.)
IV. Series
511'.6

ISBN 0 521 35923 6

Transferred to digital printing 2001

CONTENTS

Introduction

This book contains the invited papers of the conference "Algebraic combinatorics and extremal problems " held at the Université de Montréal from July 28 to August 2, 1986. This was the first time a conference focusing on these two subjects was held. The main reason for organizing such a meeting was that these apparently mutually distant parts of combinatorics are becoming more and more intertwined. This may be seen, for example, in the paper of P.J. Cameron bringing together - among other things - the representation theory of symmetric groups and the extremal problems of finite structures.

The conference proved quite useful in bringing together some of the best specialists in those two fast developing areas of combinatorics as well as others from related fields. All papers in the proceedings contain new results and most of them provide up-to-date surveys of the respective subfields accessible to non-specialists. There is a list of problems presented at the conference.

The organizers would like to use this opportunity to thank NSERC Canada for the financial support (provided by a conference grant) and the Université de Montréal and its Département de mathématiques et de statistique for hospitality. Finally we would like to thank Antoine Deza, Lucien Haddad and Marc Rosenberg for organizational help.

M. Deza, P. Frankl, I.G. Rosenberg

List of talks

Cameron	- Metric and geometric properties of permutations
S.D. Smith	- Geometric methodes in finite group theory
Fuji-Hara	- On balanced complementation of regular t-wise balanced designs
Gereb	- A common proof for Kruskal -Katona type theorems
Tichy	- Some combinatorial discrepancy problems
Anstee	- Forbidden configurations
Frankl	- Erdös-Ko-Rado type theorems
Füredi	- Families of finite sets with forbidden configurations
Kalai	- Algebraic shifting methods
Ceccherini	- A new class of Planar Pi-spaces and some related topics
Shen	- A construction of resolvable group divisible designs of block size 3
Colbourn	- Leaves and neighbourhoods in triple systems
Rosa	- Extending the concept of compatibility for triple systems
Bannai	- On extremal sets in spheres and other metric spaces
Ito	- Designs in (P and Q) - polynomial associations schemes
Wilson	- Decomposition of complete graphs
Alon	- Some recent combinatorial applications of Borsuk type theorems
Terwilliger	- A noncommutative algebra for association schemes
Hemmeter	- Cliques of quadratic forms
Labelle	- Quelques aspects de la théorie des espèces en combinatoire énumérative
Johnson	- Latin square determinants
Rosenberg	- General convolutions motivated by designs (with M.-M. Deza)
Lindström	- Matroids, algebraic and non-algebraic
J.D.S. Smith	- Infinite groups and finite quasigroups
Winkler	- Optimal realization of distance matrices
Zhou	- The chromatic difference sequence of the cartesian product of graphs
Deza	- Rigid pentagons in hypercubes
Lam	- Search for the projective plane of order 10

<u>Participants</u>

1.	N. Alon (Tel Aviv & Bell Core)	23.	J. Labelle (UQAM)
2.	R. Anstee (UBC)	24.	C. Lam (Concordia)
3.	D. Avis (McGill)	25.	B. Lindström (Stockholm)
4.	E. Bannai (Ohio State)	26.	R. Mathon (Toronto)
5.	A. Berman (Technion)	27.	A. Rosa (McMaster)
6.	I. Cahit (McMaster)	28.	I. G. Rosenberg (U.de M)
7.	P. Cameron (Oxford)	29.	G. Sabidussi (U. de M)
8.	V. Ceccherini (Rome)	30.	A. Sali (Ohio State)
9.	C. J. Colburn (Waterloo)	31.	H. Shen (Shangai)
10.	A. Deza (Neuilly)	32.	J.D.H. Smith (Iowa State)
11.	M.-M. Deza (Paris)	33.	S.D. Smith (Illinois)
12.	P. Frankl (Paris)	34.	C. Tabib (E. Montpetit)
13.	R. Fuji-hara (Tsukuba)	35.	C. Tardif (U.de M)
14.	Z. Füredi (A.T.&T. Bell Labs)	36.	P. Terwilliger (Wisconsin)
15.	M. Gereb (Harvard)	37.	R. Tichy (Vienna)
16.	A. Guelzow (Manitoba)	38.	R. Wilson (Caltech)
17.	L. Haddad (U de M)	39.	P. Winkler (Emory)
18.	J. Hemmeter (Delaware)	40.	Q. Zheng (Shangai)
19.	T. Ito (Joetsu)	41.	H. Zhou (Simon Fraser)
20.	A.V. Ivanov (Moscow)		
21.	K.W. Johnson (Penn. State)		
22.	G. Kalai (Hebrew U.)		

SOME RECENT COMBINATORIAL APPLICATIONS OF BORSUK-TYPE THEOREMS

Noga Alon
Department of Mathematics
Tel Aviv University, Ramat Aviv
Tel Aviv, Israel
and
Bell Communications Research
Morristown, NJ   07960

1. INTRODUCTION

The well known theorem of Borsuk [Bo] is the following.

Theorem 1.1 (Borsuk)

For every continuous mapping $f : S^n \rightarrow R^n$, there is a point $x \in S^n$ such that $f(x) = f(-x)$. In particular, if f is antipodal (i.e. $f(x) = -f(-x)$ for all $x \in S^n$) then there is a point of S^n which maps into the origin.

This theorem and its many generalizations have numerous applications in various branches of mathematics, including Topology, Functional Analysis, Measure Theory, Differential Equations, Approximation Theory, Geometry, Convexity and Combinatorics. An extensive list of these applications, some of which are about fifty years old, appears in [Ste].

Most combinatorial applications of Borsuk's Theorem were found during the last ten years. The best known of these is undoubtfully Lovász's ingenious proof of the Kneser conjecture. Kneser [Kn] conjectured in 1955 that if $n \geq 2r + t - 1$ and all the r-subsets of an n-element set are colored by t colors then there are two disjoint r-sets having the same color. This was proved by Lovász twenty years later in [Lo]. Shortly afterwards, Bárány [Ba] gave a charming short proof. Both proofs apply Borsuk's theorem. In [BB], Bajmóczy and Bárány deduce an interesting generalization of Radon's Theorem from Theorem 1.1. Radon's Theorem states that for any linear map f from the $(n+1)$-dimensional simplex Δ^{n+1} to the n dimensional Euclidean space R^n, there are two disjoint faces of Δ^{n+1} whose images intersect. The authors of [BB] observed that this statement, for every *continuous* map f, follows easily from Borsuk's Theorem. A more general statement was proved by Bárány, Shlosman and Szücs in [BSS]. They showed that for every prime p and every n, if $N = (p-1)(n+1)$ and $f : \Delta^N \rightarrow R^n$ is a continuous map. then there exist p pairwise disjoint faces of Δ^N, such that the intersection of all their images is nonempty. This generalizes (for prime p) a theorem of Tverberg [Tv], who proved the above for every linear map f, but without the assumption that p is a prime.

In order to establish their theorem, the authors of [BSS] proved the following interesting generalization of Borsuk's Theorem. For a prime k and for $m \geq 1$, let $X = X_{m,k}$ denote the CW-complex consisting of k disjoint copies of the $m(k-1)$ dimensional ball with an identified boundary $S^{m(k-1)-1}$. Define a free action of the cyclic group Z_k on X by defining w, the action of its generator as follows, (see [Bou], Chapter 13, for the definition of a free group action on a topological space). Represent $S^{m(k-1)-1}$ as the set of all m by k real matrices (a_{ij}) satisfying $\sum_{j=1}^{k} a_{ij} = 0$ for all $1 \leq i \leq m$ and $\sum_{i,j} a_{ij}^2 = 1$. Define now $w(a_{ij}) = (a_{i,j+1})$, where $j+1$ is reduced modulo k. Thus w just cyclically shifts the columns of a matrix representing a point of $S^{m(k-1)-1}$. Trivially, this action is free, i.e., $w(x) \neq x$ for all $x \in S^{m(k-1)-1}$. The map w is extended from $S^{m(k-1)-1}$ to $X_{m,k}$ as follows. Let (y,r,q) denote a point of X from the q-th ball with radius r and $S^{m(k-1)-1}$ - coordinate y. Then $w(y,r,q) = (wy,r,q+1)$, where $q+1$ is reduced modulo k. Since k is a prime, w defines a free Z_k action on $X = X_{m,k}$.

Theorem 1.2 ([BSS]).

For any continuous map $h:X \longrightarrow R^m$ there exists an $x \in X$, such that $h(x) = h(wx) = \cdots = h(w^{k-1}x)$.

In Sections 3 and 4, we discuss some recent combinatorial applications of this theorem.

Another interesting application of Borsuk's Theorem was given by Bárány and Lovász in [BL]. They proved that the number of vertices of any centrally symmetric simple polytope in R^n is at least 2^n (which is the number of vertices of the n-cube). Very recently, R. Stanley [Sta] proved a more general result using other algebraic methods.

There are several other cominatorial applications of Theorem 1.1, including an interesting result of Yao and Yao [YY] in computational geometry. Some of these an be found in [Bj]. In the next three sections we discuss three additional, more recent examples. The first, proved in Section 2, is the following simple result of Akiyama and the present author. The case $d=2$ of this result is a well known Putman Problem (see, e.g. [La]).

Theorem 1.3 ([AA])

Let $A_1, A_2, \ldots, A_d$ be d pairwise disjoint subsets of R^d, each containing precisely n points, and suppose that the points in $A = \bigcup_{i=1}^{d} A_i$ are in general position, (i.e., no hyperplane contains $d+1$ of the points). Then there is a partition of A into n pairwise disjoint sets $S_1, \ldots, S_n$, each containing precisely one point from each A_i, such that the n simplices conv $(S_1), \ldots,$ conv (S_n) are pairwise disjoint.

Our second example, discussed in Section 3, is the following.

Theorem 1.4 ([A1]).

Let N be an opened necklace with ka_i beads of color i, $1 \leq i \leq t$. Then it is possible to cut N in $(k-1) \cdot t$ places and partition the resulting intervals into k collections, each containing precisely a_i beads of color i, $1 \leq i \leq t$.

This theorem is best possible, and solves a problem of Goldberg and West [GW] (see also [AW]), who proved it for $k=2$. Its continuous analogue generalizes a theorem of Hobby and Rice [HR] on L_1-approximation.

In Section 4 we describe, very briefly, the proof of the following result, due to Frankl, Lovász and the present author, see [AFL].

Theorem 1.5 (The general Kneser problem)

If $n \geq (t-1)(k-1)+k \cdot r$ and all the r-subsets of an n-element set are colored by t colors then there are k pairwise disjoint r-sets having the same color.

This result is best possible and establishes a conjecture of Erdös [E], (see also [Gy]). For $k=2$ the statement of the theorem is Kneser conjecture mentioned above which was proved by Lovász. The case $r=2$ was proved by Cockayne and Lorimer [CL] and, independently, by Gyárfás [Gy]. The case $t=2$ was proved by Frankl and the present author in [AF].

Finally, in Section 5, we mention a few open problems.

2. DISJOINT SIMPLICES

As observed by Ulam, Borsuk's Theorem implies the following result, known under the self-explanatory name "the ham sandwich theorem."

Theorem 2.1

Let $\mu_1, \mu_2, \ldots, \mu_d$ be d probability measures on R^d, each absolutely continuous with respect to the usual Lebesgue measure. Then there exists a hyperplane H in R^d, which bisects all d measures, i.e., $\mu(H^-) = \mu(H^+) = \frac{1}{2}$ for all $1 \leq i \leq d$, where H^+ and H^- denote, respectively, the open positive side and the negative side of H.

Theorem 2.1 is usually deduced from Borsuk's Theorem as follows. One first shows, using measure-theoretic arguments, that for each unit vector $u \in S^{d-1}$ there is a hyperplane $H=H(u)$, perpendicular to u, with u oriented from H^- to H^+, which depends continuously on u and bisects μ_d, i.e., $\mu_d(H^-) = \mu_d(H^+)$. Next one defines a continuous function $f : S^{d-1} \rightarrow R^{d-1}$ by $f(v) = (\mu_1(H^-(v)), \ldots, \mu_{d-1}(H^-(v)))$. Since $H^+(v) = H^-(-v)$ the assertion of Theorem 2.1 now follows from that of Theorem 1.1.

We next apply the last theorem to prove the following.

Lemma 2.2

Let $A, A_1, A_2, \ldots, A_d$ be as in Theorem 1.3. Then there

$$(2.1) \qquad |H^+ \cap A_i| = \lceil n/2 \rceil \text{ and } |H^- \cap A_i| = \lceil n/2 \rceil \text{ for all } 1 \leq i \leq d.$$

(Notice that if n is odd (2.1) implies that H contains precisely one point from each A_i.)

Proof.

Replace each point $p \in A$ by a ball of radius ϵ centered in p where ϵ is small enough to guarantee that no hyperplane intersects more than d balls. Associate each ball with a uniformly distributed measure of $1/n$. For $1 \leq i \leq d$ and a (lebesgue)- measurable subset T of R^d define $\mu_i(T)$ as the total measure of balls centered at points of A_i captured by T. Clearly $\mu_1, \mu_2, \ldots, \mu_d$ are continuous probability measure. By Theorem 2.1 there exists a hyperplane H in R^d such that $\mu_i(H^+) = \mu_i(H^-) = \frac{1}{2}$ for all $1 \leq i \leq d$. If n is odd, this implies that H intersects at least one ball centered at a point of A_i. However, H cannot intersect more than d balls altogether, and thus it intersects precisely one ball centered at a point of A_i, and it must bisect these d balls. Hence, for odd n, H satisfies (2.1). If n is even, H intersects at most d balls, and by slightly rotating H we can divide the centers of these balls between H^+ and H^- as we wish, without changing the position of each other point of A with respect to H. One can easily check that this guarantees the existence of an H satisfying (2.1). $\square$

We can now prove Theorem 1.3 by induction of n. For $n=1$ the result is trivial. Assuming the result for all $n', n' < n$, let $A, A_1, A_2, ..., A_d$ be as in Theorem 1.3 and let H be a hyperplane, guaranteed by Lemma 2.2, satisfying (2.1). Put $B_i = H^+ \cap A_i$ and $C_i = H^- \cap A_i$ for $1 \leq i \leq d, B = B_1 \cup \cdots \cup B_d$ and $C = C_1 \cup \cdots \cup C_d$. By the induction hypothesis, applied to $B, B_1, \ldots, B_d$ and to $C, C_1, \ldots, C_d$ we obtain two sets S_1 and S_2 of $\lceil n/2 \rceil$ pairwise disjoint simplices each, where each simplex of S_1 contains precisely one vertex from each B_i and each simplex of S_2 contains precisely one vertex from each C_i. Clearly, all the simplices in S_1 lie in H^+ and all those in S_2 lie in H^-.

We thus obtained $2 \cdot \lceil n/2 \rceil$ pairwise non-intersecting simplices. These, together with the simplex spanned by $A_i \cap H$ if n is odd, complete the induction and the proof of the theorem. $\square$

3. SPLITTING NECKLACES

Let N be a necklace opened at the clasp with $k \cdot a_i$ beads of color i, $1 \leq i \leq t$. A *k-splitting* of the necklace is a partition of N into k parts, each consisting of a finite number of non-overlapping intervals of beads whose union captures precisely a_i beads of color i, $1 \leq i \leq t$. The *size* of the k-splitting is the number of cuts that form the intervals of the splitting. Thus, Theorem 1.4 simply asserts that every necklace with ka_i beads of color i, $1 \leq i \leq t$, has a k-splitting of size at most $(k-1) \cdot t$. One can easily check that the number $(k-1) \cdot t$ is best possible; indeed if the beads of each color appear contiguously on the opened necklace, then any k-splitting must contain at least $k-1$ cuts between the beads of each color, and hence its size is at least $(k-1) \cdot t$.

To prove Theorem 1.4 we need to formulate a continuous version of it.

Let $I = [0,1]$ be the unit interval. An *interval t-coloring* is a coloring of the points of I by t colors, such that for each $i, 1 \leq i \leq t$, the set of points colored i is (Lebesgue) measurable. Given such a coloring, a k-splitting of size r is a sequence of numbers $0 = y_0 \leq y_1 \leq \cdots \leq y_r \leq y_{r+1} = 1$ and a partition of the family of $r+1$ intervals $F = \{[y_i, y_{i+1}] : 0 \leq i \leq r\}$ into k pairwise disjoint subfamilies $F_1, \ldots, F_k$ whose union is F, such that for each $1 \leq j \leq k$ the union of the intervals in F_j captures precisely $1/k$ of the total measure of each of the t colors. Clearly, if each color appears contiguously and colors occupy disjoint intervals, the size of each k-splitting is at least $(k-1) \cdot t$. Therefore, the next theorem is best possible.

Theorem 3.1

Every interval t-coloring has a k-splitting of size $(k-1) \cdot t$

It is not difficult to check that this theorem implies Theorem 1.4. Indeed, given an opened necklace of $\sum_{i=1}^{t} ka_i = k \cdot n$ beads as in Theorem 1.4, convert it into an interval coloring by partitioning $I = [0,1]$ into $k \cdot n$ segments and coloring the j-th segment by the color of the j-th bead of the necklace. By Theorem 3.1 there is a k-splitting with at most $(k-1) \cdot t$ cuts, but these cuts need not occur at the endpoints of the $k \cdot n$ segments. One may now show, by induction on the number of "bad" cuts, that this splitting can be modified to form a k-splitting of the same size with no bad cuts, i.e., a splitting of the discrete necklace. The details are left to the reader.

Theorem 3.1 clearly follows from the following two assertions.

Proposition 3.2

Theorem 3.1 holds for every prime k.

Proposition 3.3

The validity of Theorem 3.1 for (t,k) and for (t,l) implies its validity for $(t,k \cdot l)$.

The (easy) proof of Proposition 3.3 is left to the reader. To prove Proposition 3.2 we need the following additional result from [BSS].

Put $N = (k-1) \cdot (m+1)$ and let Δ^N denote the N-dimensional simplex, i.e., $\Delta^N = \{(x_0, x_1, \ldots, x_N) \in R^{N+1}, x_i \geq 0 \text{ and } \sum_{i=0}^{N} x_i = 1\}$. The *support* of a point $x \in \Delta^N$ is the minimal face of Δ^N that contains x. Let $Y = Y_{N,k}$ denote the following CW-complex;

$$Y_{N,k} = \{(y_1, y_2, \ldots, y_k) : y_1, \ldots, y_k \in \Delta^N$$

and the supports of the $y_i - s$ are pairwise disjoint$\}$

There is an obvious free Z_k action on $Y_{N,k}$; its generator γ maps $(y_1, \ldots, y_k)$ into $(y_2, \ldots, y_k, y_1)$.

Let T and R be two topological spaces and suppose that Z_k acts freely on both. Let α and β denote the actions of the generator of Z_k on T and R, respectively. We say that a continuous mapping $f : T \to R$ is Z_k-*equivariant* if $f \circ \alpha = \beta \circ f$, (cf [Bou]. Chapter 13).

Recall that for $x \geq 0$, a topological space T is s-*connected* if for all $0 \leq \ell \leq s$, every continuous mapping of the ℓ dimensional sphere S^ℓ into T can be extended to a continuous mapping of the $\ell+1$ dimensional ball $B^{\ell+1}$ with boundary S^ℓ into T.

Lemma 3.4 [BSS].

Suppose k is a prime, $m \geq 1, N = (k-1)(m+1)$ and let $X = X_{m,k}, Y = Y_{N,k}, \mu$ and γ be as in the preceding paragraphs. Then Y is $N-k = \dim X - 1$ connected and thus there is a Z_k-equivariant map $f : X \to Y$.

We can now prove Proposition 3.2. Let k be a prime and let c be an interval t-coloring. Put $X = X_{t-1,k}$, $Y = Y_{(k-1) \cdot t, k}$ and define a continuous function $g : Y \to R^{t-1}$ as follows. Let $y = (y_1, y_2, \ldots, y_k)$ be a point of Y. Recall that each y_i is a point of Δ^N, i.e., is an $N+1$ dimensional vector with nonnegative coordinates whose sum is 1, and that the supports of the $y_i - s$ are pairwise disjoint. Put $x = (x_0, x_1, \ldots, x_N) = \frac{1}{k}(y_1 + y_2 + \cdots + y_k)$, and define a partition of $[0,1]$ into $N+1$ intervals $I_0, I_1, \ldots, I_N$, where $I_0 = [0, x_0], I_j = \left[\sum_{i=0}^{j-1} x_i, \sum_{i=0}^{j} x_i\right], 1 \leq j \leq N$. Notice that since the supports of the $y_i - s$ are pairwise disjoint, if $x_j > 0$ (i.e., the interval I_j has positive length), then there is a unique ℓ, $1 \leq \ell \leq k$ such that the j-th coordinate of y_ℓ is positive.

For $1 \leq \ell \leq k$, let F_ℓ be the family of all those $I_j - s$ such that the j-th coordinate of y_ℓ is positive. Notice that the sum of lengths of these $I_j - s$ is precisely $1/k$. For $1 \leq i \leq t-1$, define $g_i(y)$ to be the measure of the ith color in $\cup F_1$. Finally, put $g(y) = (g_1(y), g_2(y), \ldots, g_{t-1}(y))$. One can easily check that $g : Y \rightarrow R^{t-1}$ is continuous. Moreover, for $1 \leq \ell \leq k$ and $1 \leq i \leq t-1$, $g_i(\gamma^{\ell-1} y)$ is the measure of the ith color in $\cup F_\ell$. By Lemma 3.4 there exists a Z_k-equivariant map $f : X_{t-1,k} \rightarrow Y_{(t-1) \cdot k, k}$. Define $h : g \circ f : X \rightarrow R^{t-1}$. By Theorem 1.2 there is some $x \in X$ such that $h(x) = h(wx) = \cdots = h(w^{k-1} x)$. By the equivariance of f, $y = f(x)$ satisfies $g(y) = g(\gamma y) = \cdots = g(\gamma^{k-1} y)$. But this means that each of the families of intervals $F_1, F_2, \ldots, F_k$ corresponding to y captures precisely $1/k$ of the measure of each of the first $t-1$ colors. Since the total measure of each F_j is $1/k$, each F_j captures precisely $1/k$ of the measure of the last color, as well. Dividing the length 0 intervals arbitrarily between the $F_j - s$ we conclude that there is a k-splitting of size $N = (k-1) \cdot t$, as desired. This completes the proof of Proposition 3.2.

Combining the methods of this Section with a simple compactness argument one can prove the following generalization of Theorem 3.1.

Theorem 3.5

Let $\mu_1, \mu_2, \ldots, \mu_t$ be t continuous probability measures on the unit interval. Then it is possible to cut the interval in $(k-1) \cdot t$ places and partition the $(k-1) \cdot t + 1$ resulting intervals into k families $F_1, F_2, \ldots, F_k$ such that $\mu_i(\cup F_j) = 1/k$ for all $1 \leq i \leq t, 1 \leq j \leq k$. The number $(k-1) \cdot t$ is best possible.

The case $k=2$ of the last theorem is the Hobby-Rice theorem [HR] on L_1 approximation.

4. THE GENERAL KNESER PROBLEM.

The basic ideas in the proof of Theorem 1.5 are similar to those used by Lovász in [Lo], but there are several additional complications. It is first useful to reformulate Theorem 1.5 in terms of the chromatic number of a Kneser hypergraph. Let $G = G_{n,k,r}$ be the k-uniform Kneser hypergraph defined as follows. The vertices of G are all the r-subsets of $\{1,2,\ldots,n\}$, and a collection of k vertices forms an edge if the corresponding r-sets are pairwise disjoint. Theorem 1.5 is thus equivalent to the statement that if $n \geq (t-1)(k-1) + k \cdot r$ then $G_{n,k,r}$ is not t-colorable.

For any k-uniform hypergraph $H = (V,E)$, define a simplicial complex, $C(H)$ as follows; the vertices of $C(H)$ are all the $|E|k!$ ordered k-tuples $(v_1, v_2, \ldots, v_k)$ of vertices of H, where $\{v_1,\ldots,v_k\} \in E$. A set of vertices $(v_1^i, \ldots, v_k^i)_{i \in I}$ of $C(H)$ forms a simplex if there is a complete k-partite subgraph of H on the (pairwise disjoint) sets of vertices $V_1, V_2, \ldots, V_k$ such that $v_j^i \in V_j$ for all $i \in I$ and $1 \leq j \leq k$.

Theorem 1.5 now follows from the following three assertions.

Proposition 4.1

For any k-uniform hypergraph H, where k is a prime, if $C(H)$ is $(t-1)(k-1)-1$ connected, then H is not t-colorable

Proposition 4.2

$C(G_{n,k,r})$ is $(n-kr-1)$-connected.   Thus   if $n \geq (t-1)(k-1) + kr$ it is $(t-1)(k-1)-1$-connected.

Proposition 4.3

The validity of Theorem 1.1 for (r,t,k) and $(r' = (t-1)(k-1)+kr, t, k')$ implies its validity for $(r,t,k\ k')$.

Proposition 4.1 appears interesting in its own right and probably holds for every positive integer k. Its proof uses the generalization of Borsuk Theorem due to Bárány, Shlosman and Szücs, given in Theorem 1.2. Proposition 4.2 can be proved using several standard results in topology and the (easy) proof of Proposition 4.3 is purely combinatorial. The detailed proofs appear in [AFL].

Propositions 4.1 and 4.2 imply the assertion of Theorem 1.5 for every prime k. Thus, by Proposition 4.3 the theorem holds for all r,t,k.

5. <u>OPEN PROBLEMS</u>

The first obvious problem is the problem of finding pure combinatorial proofs for the problems discussed in this paper. After all, one would naturally expect that combinatorial statements about combinatorial objects should have combinatorial proofs. Such proofs are desirable, since they might shed more light on the problems. At the moment, there is no known combinatorial proof to any of the combinatorial applications of Borsuk's theorem mentioned in this paper.

Another intriguing problem is an algorithmic one. When we use Borsuk's theorem to prove the existence of a certain partition, the proof supplies no practical way for effecting such a partition. Thus, for example, one would like to find a polynomial time algorithm for finding, given an opened nacklace N with ka_i beads of color i, $1 \leq i \leq t$, a set of at most $(k-1){\cdot}t$ cuts in N and a partition of the resulting intervals into k collections, each containing precisely a_i beads of color i, $1 \leq i \leq t$. It is worth noting that we can show that the following related problem is NP-complete: Given an opened necklace N with $2a_i$ beads of color i, $1 \leq i \leq t$, and given a set of cuts of N, decide if it is possible to divide the resulting intervals into two collections, each containing precisely a_i beads of color i, $a \leq i \leq t$.

Finally we mention another problem which is related to the results of Section 3. Suppose $\mu_1, \mu_2, \ldots, \mu_t$ are t probability measures on the unit inteval I, each absolutely continuous with respect to the usual (Lebesque) measure. For a real number α, $0 \leq \alpha \leq 1$ a subset A of I is an α-share (with respect to the measures $\mu_1, \ldots, \mu_t$) if $\mu_i(A) = \alpha$ for all $1 \leq i \leq t$. We note that Liapounoff Theorem ([Li], see also [NP] and [Da]) implies that for each $\mu_1, \ldots, \mu_t$ and α as above there is an α-share A. If A is a union of a finite number s of non-overlapping intervals we define the *size* of A to be s. Otherwise, the size of A is infinity. For an integer $t \geq 1$ and $0 \leq \alpha \leq 1$ let $f(t,\alpha)$ be the smallest integer f (possibly infinity) such that for every sequence of t continuous probability measures on I there is an α-share of size at most f. Clearly $f(t,0) = f(t,1) = 1$ for all $t \geq 1$ and $f(1,\alpha) = 1$ for all $0 \leq \alpha \leq 1$. The results of Stone and Tukey [ST] easily imply that $f(2,\alpha) = 1$ for every α of the form $1/k$, k integer, and that $f(2,\alpha) = 2$ for every other α. Combining Theorem 3.5 with an appropriate construction we can show that for every two integers $t,k \geq 1$.

$$f(t,1/k) = \lfloor \frac{t{\cdot}(k-1)+1}{k} \rfloor \ .$$

This implies that $f(t,\alpha)$ is finite for every rational α. It would be interesting to decide if $f(t,\alpha)$ is finite for all possible t and α and if so, to determine or estimate this function. At the moment, we are unable to show that $f(3,\alpha)$ is finite even for a single irrational value of α.

REFERENCES

[AA] J. Akiyama and N. Alon, Disjoint simplices and geometric hypergraphs, Sugako seminar 12-85 (1985), 60, (in Japanese).

[AF] N. Alon and P. Frankl, Families in which disjoint sets have large union, Annals New York Academy of Sciences, to appear.

[Al] N. Alon, Splitting necklaces, Advances in Math, 63(1987), 247-253.

[AFL] N. Alon, P. Frankl and L. Lovász, The chromatic number of Kneser hypergraphs, Trans. AMS, 298 (1986), 359-370.

[AW] N. Alon and D. B. West, The Borsuk-Ulam Theorem and bisection of necklaces, Proc. AMS, 98 (1986), 623-628.

[Bá] I. Bárány, A short proof of Kneser's conjecture, J. Combinatorial Theory (A), 25 (1978), 325-326.

[BB] E. G. Bajomoczy and I. Bárány, On a common generalization of Borsuk's and Radon's theorems, Acta Math. Acad. Sci. Hungar. 34 (1979), 347-350.

[Bj] A. Bjorner, Topological methods, in "Handbook of Combinatorics", R. L. Graham, M. Grötshel and L. Lovász eds., North Holland, to appear.

[BL] I. Bárány and L. Lovász, Borsuk's theorem and the number of facets of centrally symmetric polytopes, Acta Math. Acad. Sci. Hungar. 40(1982), 323-329.

[Bo] K. Borsuk, Drei Sätze über die n-dimensionale euklidische sphäre, Fund. Math. 20(1933), 177-190.

[Bou] D. G. Bourgin, Modern Algebraic Topology, McMillan, New York - Collier-McMillan, London 1963.

[BSS] I. Bárány, S. B. Shlosman and A. Szücs, On a topological generalization of a theorem of Tverberg, J. London Math. Soc. (2), 23(1981), 158-164.

[CL] E. J. Cockayne and P. J. Lorimer, The Ramsey numbers for stripes, J. Austral. Math. Soc. (Ser. A) 19(1975), 252-256.

[Da] G. Darmois, Résumés exhaustifs et probléme du Nil, C. R. Acad. Sci. Paris, Vol. 222, 1946, 266-268.

[E] P. Erdös, Problems and results in combinatorial analysis, in "Coll. Internat. Th. Combinat. Rome 1973", Acad. Naz. Lincei, Rome 1976 pp. 3-17.

[Gy] A. Gyárfás, On the Ramsey number of disjoint hyperedges, J. Graph Theory, to appear.

[GW] C. H. Goldberg and D. B. West, Bisection of circle colorings, SIAM J. Alg. Discrete Methods 6 (1985), 93-106.

[HR] C. R. Hobby and J. R. Rice, A moment problem in L_1 approximation, Proc. Amer. Math. Soc. 16 (1965), 665-670.

[Kn] M. Kneser, Aufgabe 300, Jber. Deutsch. Math. Verein. 58 (1955).

[La] L. C. Larson, Problem—solving through, Springer Verlag, New York (1983), pp. 200-201.

[Li] A. Liapounoff, Sur les fonctions vecteurs completement additives, Izv. Akad. Nauk SSSR 4 (1940), 465-478.

[Lo] L. Lovász, Kneser's conjecture, chromatic number and homotopy, J. Combinatorial Th. (A) 25 (1978), 319-324.

[NP] J. Neyman and E. S. Pearson, On the problem of the most efficient tests of statistical hypotheses, Philos. Trans. Roy. Soc. London Ser. A, Vol. 231, 1932-33, 289-377.

[Pi] A. Pinkus, A simple proof of the Hobby-Rice Theorem, Proc. Amer. Math. Soc. 60 (1976), 82-84.

[Sta] R. Stanley, in preparation.

[Ste] H. Steinlein, Borsuk's antipodal theorem and its generalizations and applications; A survey, in: Coll. Sem. des Math. Sup. 95, A. Granas ed., Univ. de Montréal Press (1985), 166-235.

[ST] A. H. Stone and J. W. Tukey, Generalized sandwich theorems, Duke Math. J. 9(1942), 356-359.

[Tv] H. Tverberg, A generalization of Radon's theorem, J. London Math. Soc. 41 (1966), 123-128.

[YY] A. C. Yao and F. F. Yao, A general approach to d-dimensional geometric queries, Proc. 17th ACM STOC, ACM, Inc. Providence, R.I. (1985), 163-168.

ON EXTREMAL FINITE SETS IN THE SPHERE AND OTHER METRIC SPACES

E. Bannai
The Ohio State University, Columbus, Ohio 43210, USA

The content of this paper is, thought slightly extended, based on my expository survey talk of the same title at the Montreal meeting: Algebraic Combinatorics and Extremal Set Theory, July 27 - August 2, 1986.

The aim of this paper is to study nice finite subsets in the sphere S^d and other (nice) metric spaces. This kind of study has a long history in mathematics. Its origin is perhaps traced back to the study of regular polyhedrons in R^3 (by Platon?). In this paper, however, we restrict the scope of our discussion to the study of finite subsets which are _extremal_ from the viewpoint of Delsarte theory (which we call Algebraic Combinatorics).

This paper consists of the following four sections:

§1. Harmonics on S^d and finite sets in the sphere S^d.

§2. Combinatorics of finite sets in compact symmetric spaces of rank one.

§3. Combinatorics of finite sets in noncompact symmetric spaces of rank one.

§4. Rigid t-designs in S^d.

In §1, we give a very brief and sketchy review of the theory of finite sets in S^d (i.e., spherical codes and designs) by Delsarte, Goethals and Seidel [18], which was the starting point of the study of finite sets in topological spaces from the view point of Algebraic Combinatorics. Then we will see how this theory is generalized to other spaces, first to compact symmetric spaces of rank 1 (§2), then to noncompact symmetric spaces of rank 1 (§3), though the study in the noncompact spaces is just a beginning and yet to be developed. In §4, we discuss rigid spherical designs, which are extremal subsets of S^d from another view point.

Here I would like to offer some warnings and apologies. In this paper, I do not try to give an unbiased survey containing all the relevant results. Contrarily, I discuss only those topics to which I was personally attracted. In addition, an unusual emphasis is put on the open problems that I would like to see solved in the near future, while the results which are already theorems are mentioned only when they are useful in explaining these open problems. The content of most of this paper may not be new to experts, except possibly some presentations of new viewpoints or of new problems. The content of the last section (§4) may be new to many readers. But I admit that, partly because my study on rigid t-designs has just started very recently, the theory is in a pre-liminary stage and clearly yet to be developed.

A more detailed full account of the theories described in this paper is expected to appear as parts (i.e. Chapter 5 and Chapter 6) of the forthcoming monograph Bannai and Ito [12], which is now being prepared. I hope that this expository paper served as a handy and informal introduction to [12].

1 HARMONICS ON S^d AND FINITE SETS IN THE SPHERE S^d

Let $S^d = \{ (x_1, x_2, \ldots, x_{d+1}) \in R^{d+1} \mid x_1^2 + x_2^2 + \ldots + x_{d+1}^2 = 1\}$ be the unit sphere. The real orthogonal group $O(d+1)$ acts transitively on S^d by its natural action. S^d is identified with the homogeneous space $O(d+1)/O(d)$. It is well known that

$$L^2(S^d) = \bigoplus_{i \geq 0} \text{Harm}(i), \tag{1.1}$$

where $\text{Harm}(i)$ is the space of (complex coefficient) homogenious harmonic polynomials of degree i on R^{d+1}. (Hence dim. $\text{Harm}(i) = \binom{d+i}{i} - \binom{d+i-2}{i-2}$.) $\text{Harm}(i)$ is also the representation space of the irreducible representation ρ_i (which is called the i-th spherical representation) of $O(d+1)$.

Let X be a nonempty finite subset of S^d. For two points x and y in S^d, the distance between them $d(x,y)$ is defined as the length of the geodesic (i.e., the large circle) joining them. (Hence the diameter of S^d is π.) Clearly we have $d(x,y) = \text{arc } \cos(x,y)$, where (x,y) is the usual inner product in R^{d+1}. We define

$$A(X) = \{(x,y) \mid x,y \in X, \; x \neq y\}. \tag{1.2}$$

We say that X is an s-distance set in S^d if $|A(X)| = s$. (Set $e = |A(X)\backslash\{-1\}|$.)

The concept of t-design in S^d was defined by Delsarte-Goethals-Seidel [18] (1977).

<u>Definition</u>. A finite nonempty subset X of S^d is called a t-design if and only if

$$\frac{1}{|X|} \sum_{x \in X} f(x) = \frac{1}{|S^d|} \int_{S^d} f(x)\,d\omega(x) \tag{1.3}$$

for all polynomials $f(x) = f(x_1, x_2, \ldots, x_{d+1})$ of degree $\leq t$, where $|S^d|$ denotes the area of S^d.

There are several different but equivalent definitions of t-designs in S^d. (See [12], [18, 24], etc.) One equivalent definition is that X is a t-design in S^d if and only if

$$\sum_{x \in X} f(x) = 0 \tag{1.4}$$

for $f \in \text{Harm}(1) \oplus \text{Harm}(2) \oplus \ldots \oplus \text{Harm}(t)$. The equivalence of this definition with the previous one is easily obtained through the following decomposition theorem. Let $f(x) = f(x_1, x_2, \ldots, x_{d+1})$ be a polynomial of degree i. Then we have the following unique decomposition:

$$f(x) = \sum_{j=0}^{[\frac{i}{2}]} \|x\|^{2j} f_{i-2j}, \tag{1.5}$$

where $\|x\|^{2j} = (x_1^2 + x_2^2 + \ldots + x_{d+1}^2)^j$ and $f_{i-2j} \in \text{Harm}(i-2j)$.

The i-th spherical representation ρ_i of $O(d+1)$ is an orthogonal representation. Therefore there is an orthonormal basis in $\text{Harm}(i)\,|\,S^d$ (the domain of $\text{Harm}(i)$ restricted to S^d). The following Addition Theorem in spherical harmonics is very useful, and many proofs of the results mentioned later use this theorem.

<u>Addition Theorem on S^d</u>. Let $\{f_{i1}, f_{i2}, \ldots, f_{ih_i}\}$ (where $h_i = \dim.\text{Harm}(i) = \binom{d+i}{i} - \binom{d+i-2}{i-2})$ be an orthonormal basis of $\text{Harm}(i)\,|\,S^d$. Then we have

$$\sum_{j=1}^{h_i} f_{ij}(x)f_{ij}(y) = cQ_i((x,y)),\qquad(1.6)$$

where c is a certain constant and $Q_i(x)$ is the Gegenbauer (or ultraspherical) polynomial of degree i defined recursively by

$$Q_o(x) = 1$$

$$Q_1(x) = (d+1)x \qquad(1.7)$$

$$\lambda_{k+1}Q_{k+1}(x) = xQ_k(x) - (1-\lambda_{k-1})Q_{k-1}(x),\quad (k=1,2,\ldots)$$

with $\lambda_k = \dfrac{k}{d+2k-1}$. (Note that $h_k = Q_k(1) = \binom{d+k}{k} - \binom{d+k-2}{k-2}$, and that $Q_k(x) = \dfrac{d+2k-1}{d-1}\cdot C_k^{(d-1)/2}(x)$, (for $d \geq 2$). (See [21], [34] for $C_k^{(d-1)/2}(x)$.) We also define $R_i(x) = \sum_{j=0}^{i} Q_i(x)$, and $C_i(x) = \sum_{j=0}^{[\frac{i}{2}]} Q_{i-2j}(x)$.)

For a finite set X in S^d, let H_i be the $|X| \times h_i$ matrix (which is called the characteristic matrix) whose (x,j)-entry (with $x \in X$ and $j \in \{1,2,\ldots,h_i\}$) is $f_{ij}(x)$. Many important properties of the set X are described by using the matrices H_i through the Addition theorem. For example, X is a t-design if and only if

$$^tH_i \cdot H_j = |X|\Delta_{ij} = \begin{cases} \text{the identity martix of size } h_i\ (i=j)\\[2mm] 0 \text{ matrix } (i \neq j). \end{cases}\qquad(1.8)$$

One of the most important results in Delsarte-Goethals-Seidel [18] is summarized as follows:

Theorem 1.1. (Delsarte-Goethals-Seidel [18].) (i) If X is a t-design in S^d, then

$$|X| \geq \sum_{i=0}^{[\frac{t}{2}]} h_i = \binom{d+[\frac{t}{2}]}{[\frac{t}{2}]} + \binom{d+[\frac{t}{2}]-1}{[\frac{t}{2}]-1} = R_{[\frac{t}{2}]}(1).\qquad(1.9)$$

(ii) If X is an s-distance set in S^d, then

$$|X| \leq \sum_{i=0}^{s} h_i = \binom{d+s}{s} + \binom{d+s-1}{s-1} = R_s(1). \tag{1.10}$$

(iii) If X is a t-design as well as an s-distance set in S^d, then $t \leq 2s$. $\tag{1.11}$

(iv) If X is a subset of S^d which satisfies both the specific condition and the equality in one of the above three statements (i), (ii) and (iii), then X also satisfies the specific condition and the equality in each of the other two of (i), (ii) and (iii). If this happens, then t must be even and $t = 2s = 2e$. (We call such X a tight 2s-design.)

(v) If X is a tight 2s-design in S^d, then $A(X)$ coincides with the set of the s zeros of the polynomial $R_s(x)$.

For an antipodal X (i.e. $X = -X$), we have the following:

<u>Theorem 1.2.</u> (Delsarte-Goethals-Seidel [18].) (i) If X is antipodal and if it is a t-design in S^d, then

$$|X| \geq 2 \binom{d + [\frac{t}{2}]}{[\frac{t}{2}]} = 2 \cdot C_{[\frac{t}{2}]}(1). \tag{1.12}$$

(ii) If X is antipodal and if it is an s-distance set in S^d, then

$$|X| \leq 2\binom{d+s-1}{s-1} = 2 \cdot C_{s-1}(1) \tag{1.13}$$

(iii) If X is antipodal and if X is a t-design as well as an s-distance set in S^d, then $t \leq 2s - 1$.

(iv) If X is an antipodal subset of S^d which satisfies both the specific condition and the equality in one of the above three statements (i), (ii) and (iii), then X also satisfies the specific condition and the equality in each of the other two of (i), (ii) and (iii). If this happens, then t must be odd and $t = 2s - 1 = 2e + 1$. (We call such X a tight $(2s-1)$-design in S^d.)

(v) If X is a tight $(2s-1)$-design in S^d, then $A(X)$ concides with the set of $s(= e + 1)$ zeros of the polynomial $(x+1)C_e(x)$.

The reader will recognize that the above two theorems as well as the definition of a tight t-design are modeled to those in Q-polynomial association schemes due to Delsarte [16]. We will not discuss these theories in association schemes here. The reader is referred to Delsarte [16] and Chapter 4 of the forthcoming [12] for further details.

The above two theorems clearly show that the tight t-designs are very _extremal_ objects from the viewpoints of both t-design and s-distance set. Some examples of tight t-designs are known (see [18], [24], etc.).

Examples of tight t-designs in S^d

(a) Tight t-design in S^1 is the set of $t + 1$ vertices of a regular $(t+1)$-gon.

(b) Tight 1-design in S^d is a pair of 2 antipodal points.

(c) Tight 2-design in S^d is the set of $d + 2$ vertices of a regular simplex.

(d) Tight 3-design in S^d is the set of $2d$ vertices of a generalized regular octahedron.

(e) There are two known tight 4-designs in S^d $(d \geq 2)$: One is $|X| = 27$ for $d = 5$; the other is $|X| = 275$ for $d = 21$.

(f) There are three known tight 5-designs in S^d $(d \geq 2)$: One is $|X| = 12$ for $d = 2$; the others are $|X| = 56$ for $d = 6$ and $|X| = 552$ for $d = 22$.

(g) There are two known tight 7-designs in S^d $(d \geq 2)$: One is $|X| = 240$ for $d = 7$; the other is $|X| = 4600$ for $d = 22$.

The following theorem says that tight t-designs in S^d $(d \geq 2)$ do not exist for large t. The proof of it uses the Lloyd type theorem ((v) in Theorem 1.1 and Theorem 1.2) that states that the zeros of $R(x)$ and $C(x)$ must be all rational numbers (if X is tight), and the number theoretical study that shows that that does not occur for large t.

Theorem 1.3. (Bannai-Damerell [6, 7]) (i) For $d \geq 2$, if $e \geq 3$ then there is no tight 2e-design in S^d.

(ii) For $d \geq 2$, if there is a tight $(2e+1)$-design in S^d, then either $e \leq 3$, or $e = 5$ and $d = 23$ (and $|X| = 196560$). (The uniqueness of the tight 11-design in S^{23} $(|X| = 196560)$ was shown by Bannai-Sloane [13]).

Here we mention several open questions which I believe can

be solvable in the near future.

Problem 1.1. Determine tight t-designs in S^d for $t = 4$, 5 and 7. Remark: These three values of t are the only open cases. Note that if $d \geq 2$ if there is a tight t-design in S^d, then we have a necessary condition which comes from a Pell equation, obtained by the fact that the zeros of $R_2(x)$ (for $t = 4$), $C_2(x)$ (for $t = 5$), $C_3(x)$ (for $t = 7$) are all rational numbers. For example, $d + 4$ must be a square of an integer for $t = 4$. But what does happen if $d + 4$ is a square (for $t = 4$)?

Problem 1.2. Determine non-antipodal X in S^d with $t = 2s - 1$. (Assume that t is not too small if necessary.) Remark: Perhaps it should be possible also to treat the case $t = 2s - 2$ (with X either antipodal or non-antipodal, assuming t is not too small if necessary). As is well known, if $t \geq 2s - 2$, then X has a structure of Q-polynomial association scheme of class s, so they are expected to exist very rarely.

Problem 1.3. Special case of Problem 1.2. If we assume furthermore in Problem 1.2 that X is the set of minimal vectors of an even unimodular lattice in R^{d+1}, then what can we say? (Is there any advantage of assuming this extra assumption?)

Problem 1.4. Determine $2s$-design in S^d of cardinality $R_s(1) + 1$. Then determine $2s$-design with cardinality $R_s(1) + 2$, and so on, assuming t is not too small if necessary. Also, similar problems for antipodal $(2e+1)$-designs in S^d with cardinality $2C_e(1) + 1$, etc.

Problem 1.5. Determine s-distance set X in S^d of cardinality $R_s(1) - 1$. Then determine s-distance set of cardinality $R_s(1) - 2$, and so on. Also, similar problems for antipodal s-distance sets in S^d with cardinality $2 \cdot C_e(1) - 1$, etc.

2   COMBINATORICS OF FINITE SETS IN COMPACT SYMMETRIC SPACES OF RANK 1

Another important paper [17] by Delsart-Goethals-Seidel (which precedes [18]) treats the configurations of lines (through the origin) in R^n and C^n. Since the lines in R^n and C^n are regarded as the points in the projective spaces over R and C, we can regard

the paper studying finite subsets in these projective spaces. The theory
here goes quite parallel to the theory in S^d (which we have outlined
in the previous section). Actually, there is an important class of
spaces (i.e., compact symmetric spaces of rank 1) to which this theory is
generalized very naturally and very straightforwardly. This is what we
want to discuss in this section.

Compact symmetric spaces of rank 1 are classified and their
properties are well known. Let M be a compact symmetric space of rank
1. It must be one of the following spaces, which are obtained as the
homegeneous spaces G/H.

$$
\begin{aligned}
&\text{(a)} \quad M = S^d = SO(d+1)/SO(d), \quad d = 1,2,3,\ldots \\
&\text{(b)} \quad M = P^d(R) = SO(d+1)/O(d), \quad d = 2,3,4,\ldots \\
&\text{(c)} \quad M = P^d(C) = SU(\tfrac{1}{2}d+1)/U(\tfrac{1}{2}d), \quad d = 4,6,8,\ldots \\
&\text{(d)} \quad M = P^d(H) = Sp(\tfrac{1}{4}d+1)/S_p(1) \times S_p(\tfrac{1}{4}d), \quad d = 8,12,16,\ldots \\
&\text{(e)} \quad M = P^{16}(0) = F_4/Spin(9).
\end{aligned}
\tag{2.1}
$$

(Note that d stands for the topological dimension of the space M, and
that $P^1(R) \cong S^1$, $P^2(C) \cong S^2$, $P^4(H) \cong S^4$ and $P^8(0) \cong S^8$.)

A detailed explanation of these spaces (from the viewpoint of
combinatorics) will be found in Hoggar [26]. (See also [25], [37] for
the properties of these spaces from a general viewpoint.)

Since M is a Riemannian manifold, using the natural distance, the concept of s-distance set in M is defined naturally. As in
S^d, we can also consider the set of inner products A(X). Take $P^d(K)$,
the projective space over K, K being one of R, C and H. (The case
K = 0 needs a modification, cf. [26], [27].) Then there is a natural
vector space over K (of an appropriate dimension) on which G acts
naturally. Also, there exists a natural (hermitian) inner product, and
the points in $P^d(K)$ are represented by vectors of the unit norm (with
respect to the inner product). Using this identification the distance
d(x,y) between two points x and y in $P^d(K)$ is given by:

$$
d(x,y) = \text{arc cos}\,|(x,y)|^2.
\tag{2.2}
$$

For a finite subset X of M, we define (as before), A(X) =
$\{|(x,y)|^2 | (x,y) \in X, x \neq y\}$ (so $A \subset [0,1)$). (Set $|A(X)| = s$ and
$|A(X)\backslash\{0\}| = e$. So $\epsilon = s - e$ is 0 or 1 according as $0 \in A(X)$ or

not. Since M is a symmetric space of rank 1, we get (this is the reason why we call rank 1) that

$$L^2(M) = \bigoplus_{i \geq 0} V_i, \tag{2.4}$$

where the irreducible G-spaces V_i (which are finite dimensional spaces) have the natural ordering indexed by non-negative integers.

<u>Definition</u>. A finite set X of M is called a t-design in M if and only if

$$\sum_{x \in X} f(x) = 0 \quad \text{for all} \quad f(x) \in V_1 \oplus V_2 \oplus \ldots \oplus V_t. \tag{2.5}$$

(This is equivalent to the condition that $\frac{1}{|X|} \sum_{x \in X} f(x) = \frac{1}{|M|} \int_M f(x)\, d\omega(x)$ for all $f(x) \in V_0 \oplus V_1 \oplus \ldots \oplus V_t$.

A result analogous to Theorem 1.1 is then obtained exactly the same way as Theorem 1.1 was proved.

<u>Theorem 2.1</u>. (Cf. [26], [33]) Let M be a projective space $P^d(K)$, with K being one of R, C, H and O.

(i)   If X is a t-design in M, then $|X| \geq \sum_{i=0}^{[\frac{t}{2}]} h_i = R_{[\frac{t}{2}]}^0 (1).$ (2.6)

(ii)   If X is an s-distance set in M, then $|X| \leq \sum_{i=0}^{s} h_i = R_s^0(1).$ (2.7)

(iii)   If X is a t-design as well as an s-distance set in M, then $t \leq 2s$.

(iv)   Exactly the same statement as in Theorem 1.1(iv) holds when we replace S^d by M. (We call such X a tight 2s-design.)

(v)   If X is a tight 2s-design in M, then A(X) coincides with the set of the s zeros of the polynomials $R_s^0(x)$.

Here we use the following notation: $m = \frac{1}{2}(K:R)$, $N = m(\frac{d}{2m} + 1)$, $\varepsilon = 0$ or 1. $(= s - e)$.

$$Q_k^\varepsilon(x) = \frac{(N)_{2k+\varepsilon}}{(m)_{k+\varepsilon}} \sum_{i=0}^{k} (-1)^i \binom{k}{i} \frac{{}_i(k+m+\varepsilon-1)}{{}_i(2k+N+\varepsilon-2)} x^{k-i} \qquad (2.8)$$

and

$$R_k^\varepsilon(x) = \sum_{j=0}^{k} Q_j^\varepsilon(x), \qquad (2.8)\text{bis}$$

where $_0(q) = 1 = (q)_0$ and $_a(q) = q(q-1) \ldots (q-a+1)$, $(q_a) = q(q+1) \ldots (q+a-1)$ for $a \geq 1$.

Note that $Q_i^\varepsilon(x)$ ($\varepsilon = 0$ or 1) here are different from the $Q_i(x)$ (for S^d) in the previous section, although all of them are Jacobi polynomials of certain (different) parameters. Also, note that

$$Q_k^\varepsilon(1) = (N)_{k+\varepsilon-1}(N-m)_k(2k+N+\varepsilon-1)/(m)_{k+\varepsilon} \cdot k!$$

and

$$R_k^\varepsilon(1) = (N)_{k+\varepsilon} \cdot (N-m+1)_k/(m)_{k+\varepsilon} \cdot k!,$$

and $h_i = \dim V_i = Q_i^0(1)$. (See [26], [23] for the details).

Here we remark that G acts on V_i as a unitary representation, and V_i has an orthonormal basis $\{f_{i1}, f_{i2}, \ldots, f_{ih_i}\}$. The addition theorem is also obtained on M and that plays an important role in the proof of the above theorem (as well as many other theorems).

Addition Theorem on M. For $x, y \in M$, we have

$$\sum_{j=1}^{h_i} f_{ij}(x)\overline{f}_{ij}(y) = c \cdot Q_i^0(|(x,y)|^2), \qquad (2.9)$$

where c is a constant and $Q_i^0(x)$ is a Jacobi polynomial (explicitly known for each M) described in (2.8). (See [30] etc., for the proof of the addition theorem on M).

A result analoguous to Theorem 1.2 is also obtained, and is stated as follows.

Theorem 2.2. (Cf. [20], [26], [33]) Let X be a finite set in $M = P^d(K)$, where K is one of R, C, H and O.

(i) If X is a t-design in M, then $|X| \geq \sum_{i=0}^{[\frac{t-1}{2}]} h_i^*$
$(= R_{[\frac{t-1}{2}]}^1 (1))$, where h_i^* are certain well defined numbers for each M, (i.e., $h_i^* = Q_i^1(1)$).

(ii) If X is an s-distance set in M and if $0 \in A(X)$, then $X \geq \sum_{i=0}^{e} h_i^* (= R_e^1(1))$.

(iii) If X is a t-design in M and if X is also an s-distance set with $0 \in A(X)$, then we have $t \leq 2s - 1 (= 2e + 1)$.

(iv) If a finite set X in M satisfies both the specific condition and the equality in one of the above three statements (i), (ii), and (iii), then X also satisfies the specific condition and the equality in each of the other two of (i), (ii), and (iii). If this happens, then t must be odd and $t = 2s - 1 = 2e + 1$. (We call such X a tight $(2e + 1)$-design in M.)

(v) If X is a tight $(2e + 1)$-design in M, then $A(X)$ coincides with the set of $e + 1$ zeros $xR_e^1(x)$.

Again, tight t-designs are the most extremal subsets from the view points of both s-distance sets and t-designs. Several interesting examples of tight t-designs in $P^d(K)$ are described in [26].

We have recently completed the proof of the following theorem.

Theorem 2.3. (Bannai and Hoggar [8,9,10]) If there exists a tight t-design in $M = P^d(K)$ (with $d \geq 2$), then $t \leq 5$.

(Note: The result for $K = R$ is due to Bannai and Damerell [7]. The result for $K = 0$ is due to Hoggar [27].)

The proof of this theorem uses Lloyd type theorem (originally designed for the (non)-existence of perfect codes). Namely, if X is a tight t-design in $P^d(K)$, then we can show that the set $A(X)$ must be identical with the zeros of a Jacobi polynomial $R_s^\varepsilon(x)$ (if $t = 2s$)   or $xR_e^\varepsilon(x)$ (if $t = 2s - 1$) (described in (2.8) and (2.8)bis), whose co-efficients are all rational numbers. Since we can easily see that all the h_i's   (or h_i^*'s) are distinct, we can conclude that all the zeros of this polynomial are rational numbers. Then after a very involved argument which partly uses the method in [22] and models itself on the proof in [7], together with various ad-hoc method, we can solve this number theortical problem completely for $t > 5$. (See [10] etc. for the details of the argument.)

We conclude this section by giving several open problems.

<u>Problem 2.1.</u>   Determine tight 4- and tight 5-designs in $M = P^d(K)$.   Remarks:

(i)   There is a tight 5-design in $P^{23}(R)$ with $|X| = 98,$ 280.   (This is obtained from the tight 11-design in S^{23} by identifying antipodal points.)

(ii)   There is a tight 5-design in $P^{16}(0)$ with $|X| = 819.$

(iii)   No examples of tight 4-designs in $M = P^d(K)$ $(d \geq 2)$ are known, and they are not likely to exist.

(iv)   It is known that each of the above tight 5-designs in (i) and (ii) is unique in $P^d(K)$ for the relevant   d and   K.   [(i) is immediately obtained from Bannai-Sloane [13]. The uniqueness of (ii) is reduced to the uniqueness of the generalized hexagon of type ${}^3D_4(2)$ (I owe this remark to A. Gardiner), and the uniqueness of such generalized hexagon was proved by A. Cohen.]

(v)   There are some necessary conditions for the existence of tight 4- and 5-designs which come from the Lloyd type theorem (cf. Problem 1.1.)

<u>Problem 2.2.</u>   Study the obvious problems corresponding to Problems 1.2, 1.4 and 1.5 for $M = P^d(K)$.

<u>Problem 2.3.</u>   If $t \geq 2s - 2$, then   X   in   M   has a structure of the Q-polynomial association scheme. It is known that such association scheme must satisfy certain extra conditions on its parameters. (Hoggar [28] studied this for the special case $t = s = 2$ and $A(x) = \{0,\alpha\}$). What can we say about the converse? For example, can one determine when

a strongly regular graph (i.e., symmetric association scheme of class 2) is realized as a 2-design and 2-distance set in $P^d(K)$ for given K. Does this realizability depend only on the parameters of the strongly regular graph? In other words, are there two different strongly regular graphs with the identical parameters, one is realized in $P^d(K)$ and the other is not realized in $P^d(K)$?

Problem 2.4. (As I am not very familiar with differential geometry, I am afraid that some of my explanations below might not be free from some technical inaccuracy. So please regard that Problem 2.4 includes the following problem: clarify what I am trying to say.) A concept (which was called Delsarte space) was introduced by Neumainer [33] as a concept which includes both Q-polynomial association scheme and compact symmetric spaces of rank 1. It is true that all the algebraic properties of both of these two types of spaces are derived from the axioms given by Neumaier [33]. But I think that there is still room for the discussion about what should be the best definition of the Delsarte space.

1. What is the best definition of the Delsarte space?
2. Either for that definition or for the Neumaier's definition, can one determine Delsarte spaces which are compact Riemannian manifolds? That is, are there examples other than compact symmetric spaces of rank 1.
3. The definition of Delsarte space in [33] _does_ depend on the properties of the embedding of the space into R^d. I think it would be better if this could be avoided. Can it be possible? Is there any nice differential geometric definition of the Delsarte space (e.g., using some tensors)?
4. In [36] Terwilliger gives a characterization of the Q-polynomial property of association schemes, which also depends on the property of the embedding of the association scheme in R^d. Can one get a characterization (definition) of compact symmetric spaces of rank 1 in the line of Terwilliger [36]? Assuming that this is done, find what relations exist between Neumaier's Delsarte space and this definition.

5. Among compact Riemannian manifolds, what kind of characterizations of compact symmetric spaces of rank 1 are possible? Combinatorial characterization, i.e., something like using the definition similar to distance-regularity rather than distance-transitivity, is preferred. Also the characterizations by the spectrum (i.e., the eigenvalues of Laplace-Beltrami operator, and so on) are preferred. (We do not want to assume the property of isometries of M.)

6. Let M be a compact Riemannian manifold. Let $0 = \lambda_o < \lambda_1 < \lambda_2 < \ldots$ be the distinct eigenvalues of the Laplace-Beltrami operator (see e.g. [15]), and let V_i be the eigenspace corresponding to the eigenvalue λ_i (V_i's are finite dimensional subspaces of $L^2(M)$). Roughly speaking, suppose that $f \cdot g$ with $f \in V_1$ and $g \in V_i$ is in $V_{i-1} \oplus V_i \oplus V_{i+1}$ (for all i) and the projection of $f \cdot g$ to each of V_j ($j = i-1, i, i+1$) ranges over uniformly when f and g range over V_1 and V_i. Then, does this property characterize the compact symmetric spaces of rank 1?

(Many other interesting open problems can be found in [26], [33].)

3 COMBINATORICS OF FINITE SUBSETS IN NON-COMPACT SYMMETRIC SPACES OF RANK ONE

The most interesting spaces other than the sphere and the projective spaces would be the real Euclidean space and the hyperbolic spaces. There is a well defined class of spaces, namely the class of non-compact symmetric spaces of rank 1, which includes both of these spaces. Namely, we have the following classification. Let M be a non-compact symmetric spaces of rank 1. Then M is one of the following spaces which are realized as the homogeneous space G/H:

(0) $M = R^d = E(d)/O(d) = R^d O(d)/O(d)$, with $d = 1,2,3,\ldots$

(1) $M = H^d(R) = SO^1(d+1)/SO(d)$, with $d = 2,3,4$

(2) $M = H^d(C) = SU^1(\frac{1}{2}d+1)/U(\frac{1}{2}d)$, with $d = 4,6,8,\ldots$

(3) $M = H^d(H) = Sp^1(\frac{1}{4}d+1)/Sp(1) \times Sp(\frac{1}{4}d)$, with $d = 8,12,16,\ldots$

(4) $M = H^{16}(0) = F_4^*/\mathrm{Spin}(9)$.

(For the details of these spaces, see e.g. Helgason [25] and/ or Wolf [37].)

Since M is non-compact here, there are many differences from compact symmetric spaces. However, since M is a Riemannian manifold, there exists the natural distance, and so the concept of s-distance set in M is obviously defined. 2-distance sets in R^d were first studied by Larman-Rogers-Seidel [31], and it was shown that if X is a 2-distance set in R^d, then $|X| \leq \dfrac{(d+1)(d+4)}{2}$. (The proof uses elementary harmonic analysis on S^d, in particular the formula (1.5) mentioned in §1.) This bound was not sharp, and we now obtain the following better bound for any s (by analyzing and improving the method of proof in [31]).

Theorem 3.1.   (Cf. [14], [4]) Let X be an s-distance set in R^d, then $X \leq \binom{d+s}{s}$.

(We don't know any example of s-distance set X whose cardinality attains the bound $\binom{d+s}{s}$ except for either $s = 1$ or $d = 1$.)

For the real hyperbolic space $H^d(R)$, we have the following:

Theorem 3.2.   (Cf. [14], [5]) Let X be an s-distance set in $H^d(R)$, then we have $|X| \leq \binom{d+s}{s}$. (This bound does not seem to be sharp except for either $s = 1$ or $d = 1$.)

Proof of Theorem 3.2 given in [5] may be interesting, because we use the following addition theorem on $H^d(R)$, which seems to have been unnoticed before.

First, let us fix some notation.

Let $M = H^d(R)$ be represented in R^{d+1} by

$$H^d(R) = \{(x_1, x_2, \ldots, x_{d+1}) \in R^{d+1} \mid x_1^2 - x_2^2 - \cdots - x_{d+1}^2 = 1\}$$

with $d(x,y) = \mathrm{arc\,cosh}\langle x,y \rangle = \mathrm{arc\,cosh}(x_1 y_1 - x_2 y_2 - \cdots - x_{d+1} y_{d+1})$ for $x = (x_1, \ldots, x_{d+1})$ and $y = (y_1, \ldots, y_{d+1}) \in H^d(R)$. Let

$$\Delta = \frac{\partial^2}{\partial x_1^2} - \frac{\partial^2}{\partial x_2^2} - \cdots - \frac{\partial^2}{\partial x_{d+1}^2}.$$

Define $\mathrm{Harm}(i) = (\mathrm{Ker}\,\Delta) \cap \mathrm{Hom}(i)$, where $\mathrm{Hom}(i)$ denotes the space of homogeneous polynomials of degree i in $x_1, x_2, \ldots, x_{d+1}$. Then

$$\text{dim. Harm}(i) = \binom{d+i}{i} - \binom{d+i-2}{i-2}$$

$$= \binom{d+i-1}{i-1} + \binom{d+i-2}{i-2}$$

<u>Addition Theorem on $H^d(R)$</u>. ([5]) $\text{Harm}(i)$ has a basis $\{f_{i1}, \ldots, f_{i\mu_i}, g_{i1}, \ldots, g_{i\nu_i}\}$ with

$$\mu_i = \binom{d+i-1}{i-1} \quad \text{and} \quad \nu_i = \binom{d+i-2}{i-2} \quad \text{for even } i, \quad \text{and}$$

$$\mu_i = \binom{d+i-2}{i-2} \quad \text{and} \quad \nu_i = \binom{d+i-1}{i-1} \quad \text{for odd } i,$$

such that

$$\sum_{i=1}^{\mu_i} f_{ij}(x) f_{ij}(y) - \sum_{j=1}^{\nu_i} g_{ij}(x) g_{ij}(y) = c \cdot Q_i(\langle x, y \rangle),$$

for $x, y \in H^d(R)$ where c is a constant and $Q_i(x)$ is the Gegenbauer polynomial of degree i (defined in (1.7)).

Using this addition theorem, Theorem 3.1 is proved in a similar manner as the upper bound of the cardinality of an s-distance set in S^d (i.e., Theorem 1.1(ii)) was obtained by using the addition formula on S^d together with the formula (1.8). (See [5] for the details.)

When we try to define a concept of t-design for finite sets in M, the noncompactness of M gives us a real trouble. One reason of difficulty is that the integral of a constant function on M is not of finite value. I have been trying to find a reasonable definition of t-design in R^d or $H^d(R)$ for several years, but I am not yet successful. Since a (nonbounded) noncompact space cannot be approximated by any finite set, this intension might be deemed to fail. But I still cannot help hoping this might be possible. I feel that the addition theorem we got for $H^d(R)$ should be relevant to the possible definition of t-design in $H^d(R)$, and it is lucky that we have the addition theorem on $H^d(R)$ despite the difficulty of handling non-compact spaces.

Take $M = R^d$. The structure of $L^2(R^d)$ is well known and well studied (cf. e.g. [32]). But the difficulty is that the E(d)-

invariant irreducible (unitary) subspaces appearing in $L^2(R^d)$ are not finite dimensional and that they are not parameterized by the non-negative integers, (they are parameterized by a continuous parameters). Nonetheless, the following is one possible candidate of the definition of t-design in R^d (although the homogenuity of R^d with respect to the translation is broken down when we use this definition).

A candidate of definition of t-design in R^d. A finite set X in R^d is called a "t-design" if we have

$$\sum_{x \in X} e^{-\|x\|^2} f(x) / \sum_{x \in X} e^{-\|x\|^2} = \int_{R^d} e^{-\|x\|^2} f(x)\, dx / \int_{R^d} e^{-\|x\|^2}\, dx,$$

for all polynomials $f(x) = f(x_1, x_2, \ldots, x_d)$ of degree $\leq t$. (But I am not sure whether this is a reasonable definition of t-design, or whether we can get many interesting results out of this definition.)

I believe that it would be most desirable if we can find the definition of t-design in R^d (or $M = H^d(K)$) for which something like Theorem 1.1 (or 1.2, 2.1, or 2.2) holds.

We conclude this section by giving several open problems.

Problem 3.1. (1) Determine whether there exists an s-distance set X in R^d with $|X| = \binom{d+s}{s}$ for $s \geq 2$ and $d \geq 2$.

(1)' If there is such a set, then determine all the s-distance sets in R^d with $|X| = \binom{d+s}{s}$.

(1)" If there is not such set, then determine what is the natural new upper bound for the cardinality of an s-distance set X in R^d. Then determine the s-distance sets in R^d which attain the new upper bound.

(2) Assuming that the natural upper bound is determined in (1), find a reasonable definition of t-design for finite subsets in R^d for which the result like Theorem 1.1 is obtained (if necessary by modifying the definition of s-distance set so that a theorem like Theorem 1.1 holds).

Problem 3.2. Try the same problem as Problem 3.1 for s-distance sets X in $H^d(R)$.

Problem 3.3. (1) Find an addition theorem as Theorem 3.3 for $M = H^d(K)$ with the $K = C, H, O$.

(2) Using the addition formula obtained in (1), find an

upper bound for the cardinality of an s-distance set in $M = H^d(K)$, with the $K = C$, H, O.

(3)   Then do the same problem as Problem 3.2 for s-distance set X in $H^d(K)$, with the $K = C$, H, O.

<u>Problem 3.4</u>. Since R^d is a limiting case of $H^d(R)$, it may not be a big surprise that we have the same upper bound $\binom{d+s}{s}$ for both cases. Can one find an addition theorem on R^d which is obtained as a limiting case of the addition Theorem 3.3?

<u>Problem 3.5</u>. In [19] Deza and Frankl studied finite sets X in R^d such that $|\{(x,y) \mid x,y \in X, x \neq y\}| = s$ (where (x,y) is the usual inner product on R^d), and proved that $|X| \leq \binom{d+s}{s}$.

When is the upper bound attained? Can one find any connection of such sets with s-distance sets in R^d? (Although this is not the exact one, I think that this type of modification of s-distance set in R^d might produce a result like Theorem 1.1 combined with the proper definition of t-design in R^d. So we should explore many possibilities without fearing to fail.)

4   <u>RIGID t-DESIGNS IN S^d</u>

In the previous sections, we have seen that the tight t-designs are the most extremal finite sets from the viewpoints of both t-designs and s-distance sets, and so far we discussed mostly the topics related to tight t-designs. In this section, another type of extremal finite sets in S^d are introduced. That is the concept of rigid t-design in S^d. I believe this is an interesting concept, and that the classification problem of all the rigid t-designs in S^d is a very challenging important problem.

First let us look at t-designs in S^1.

In the circle S^1, it is easy to see that the $k + 1$ vertices of a regular $(k+1)$-gon with $k \geq t$ (embedded in S^1) form a t-design. In [29] Y. Hong proved the following results for spherical t-designs in S^1.

    (i)   If $|X| \leq 2t + 1$, then X must be a regular $(k+1)$-gon with $t \leq k \leq 2t$,

    (ii)   If $|X| = 2t + 2$, then X must be a union of two regular $(t+1)$-gons,

(iii)   If $|X| \geq 2t + 3$, then there are infinitely many non-group type spherical t-designs, where group type means a union of regular $(k_i + 1)$-gons with $k_i \geq t$.

This result of Hong suggested the existence of spherical t-designs in abundance. The existence of spherical t-designs in S^d for any t and d was proved by Seymour-Zaslavsky [34] in a very general context.

In the above result of Hong, the t-designs X with $|X| \leq 2t + 1$ are markedly different from those with $|X| \geq 2t + 2$. Namely, those with $|X| \leq 2t + 1$ are all not deformable, but those with $|X| \geq 2t + 2$ are all deformable. While those t-designs constructed (i.e., shown to exist) by Seymour-Zaslavsky [34] are all deformable from the nature of their proof, the tight t-designs are not deformable from their nature. Partly motivated by these observations, I formulated the following concept of rigid t-designs in S^d (or called rigid spherical t-designs in S^d), and asked whether there are many such examples or not (cf. [1,2]).

<u>Definition</u>. We call $X = \{\vec{x}_1, \vec{x}_2, \ldots, \vec{x}_n\}$ is a <u>non-rigid</u> (or deformable) spherical t-design in S^d, if for any given $\varepsilon > 0$ there exists another spherical t-design $X' = \{\vec{x}'_1, \vec{x}'_2, \ldots, \vec{x}'_n\}$ such that $\|\vec{x}_i - \vec{x}'_i\| < \varepsilon$ (for $1 \leq i \leq n$) and there exists no orthogonal transformation 0 on R^{d+1} with $0\vec{x}_i = \vec{x}'_i$ $(1 \leq i \leq n)$.

<u>Definition</u>. We call X a <u>rigid</u> (or non-deformable) if it is not non-rigid.

I was eventually led to conceive the following conjectures.

<u>Conjecture 1</u>. For each fixed pair of t and d, if $|X|$ is sufficiently large (i.e., greater than a certain number $f(t,d)$ depending only on t and d), then X is non-rigid.

<u>Conjecture 2</u>. For each fixed pair of t and d, there are only finitely many rigid spherical t-designs up to orthogonal transformations.

<u>Remarks</u>.

(i)   Conjecture 2 implies Conjecture 1.

(ii)   Tight t-designs are examples of rigid spherical t-designs. There are some other known rigid spherical t-designs, but they are very rare (at least as far as I am aware of at the present time).

(iii) It seems that a rigid spherical t-design may represent a stable state (from the viewpoint of moments) of finitely many particles in S^d. So, the classification problem (if at all possible) may be an interesting question from the viewpoint of physics. This question will be even more interesting if Conjecture 2 is proved to be true.

Recently, I proved the complete result for S^1 concerning Conjecture 1 and Conjecture 2.

<u>Theorem 4.1.</u> ([3]) If X is a rigid spherical t-design in S^1, then X consists of the $k+1$ vertices of a regular $(k+1)$-gon with $t \leq k \leq 2t$.

The proof of Theorem 4.1 is not very difficult. The implicit function theorem plays a key role. Here we will explain the basic idea of the proof in a general context. (We prove that a t-design X in S^1 is not rigid if $|X| \geq 2t + 2$. Then Hong's result [29] implies the rest.)

Generally, let X be a t-design in S^d, and let $X = \{\vec{x}_j \mid j = 1,2,\ldots,|X|\}$. Let $(\xi_1,\xi_2,\ldots,\xi_d)$ be a local coordinate of S^d, and let $\vec{x}_j = (\xi_{1j},\xi_{2j},\ldots,\xi_{dj})$. Let $\{f_{i1},f_{i2},\ldots,f_{ih_i}\}$ be a basis of $\mathrm{Harm}(i)$.

Now, let us consider the Jacobian matrix J of size $(h_1 + h_2 + \ldots + h_t) \times |X|d$ whose rows are indexed by the basis of $\mathrm{Harm}(1) \oplus \mathrm{Harm}(2) \oplus \ldots \oplus \mathrm{Harm}(t)$, whose columns are indexed by the pairs of an element of X and a local coordinate ξ_k $(k \in \{1,2,\ldots,d\})$, and whose $(f_{i\ell},(\vec{x}_j,k))$ entry is given by $\dfrac{\partial f_{i\ell}}{\partial \xi_k}(\xi_{1j},\xi_{2j},\ldots,\xi_{dj})$.

So,

$$
J = \left.
\begin{array}{c}
h_1 \left\{ \rule{0pt}{40pt}\right. \\[10pt]
h_i \left\{
\begin{array}{c}
f_{i1} \\
f_{i\ell} \\
f_{ih_i}
\end{array}
\right. \\[10pt]
h_t \left\{ \rule{0pt}{30pt}\right.
\end{array}
\right(
\begin{array}{c}
\overbrace{\hspace{6cm}}^{\displaystyle d} \\
\xi_{1j} \qquad \xi_{kj} \qquad \xi_{dj} \\[10pt]
\text{— — — — — — }\dfrac{\partial f_{i\ell}}{\partial \xi_k}(\xi_{1j},\xi_{2j},\ldots,\xi_{dj})
\end{array}
\right)
$$

Suppose that J has the maximal possible rank $h_1 + h_2 + \ldots + h_t$ and that $|X| d - \dim O(d+1) > h_1 + h_2 + \ldots + h_t$. Then X is <u>not</u> rigid. Because, by the implicit function theorem, a slight change of certain $|X| d - (h_1 + h_2 + \ldots + h_t)$ variables among ξ_{jk} $(1 \leq j \leq |X|$ and $1 \leq k \leq d)$ determines the (slight) change of the rest of the variables ξ_{jk} and we get a new t-design X'. So, if these free parameters are more than the $\dim O(d+1)$, then some of such obtained X' are not transformed to X by an orthogonal transformation, which shows that X is not rigid.

In S^1 - case, we take θ as a local coordinate. Since $h_i = 2$ for all $i \geq 1$, and since $\mathrm{Harm}(i) = \{f_{i1}, f_{i2}\} = \{\cos i\theta, \sin i\theta\}$, we are reduced to show that the matrix J (given below) has the maximal possible rank $2t$, if $|X| \geq 2t + 1$:

$$
J = \begin{pmatrix}
\cos\theta_1 & \cos\theta_2 & \cdots & \cos\theta_{|X|} \\
\sin\theta_1 & \sin\theta_2 & \cdots & \sin\theta_{|X|} \\
\cos 2\theta_1 & \cos 2\theta_2 & \cdots & \cos 2\theta_{|X|} \\
\sin 2\theta_1 & \sin 2\theta_2 & \cdots & \sin 2\theta_{|X|} \\
\vdots & & & \vdots \\
\cos t\theta_1 & \cos t\theta_2 & \cdots & \cos t\theta_{|X|} \\
\sin t\theta_1 & \sin t\theta_2 & \cdots & \sin t\theta_{|X|}
\end{pmatrix}
$$

This is actually proved (cf. [3]) without much difficulty, and since
$\dim 0(d+1) = 1$ for $d = 1$, we get Theorem 4.1.

In the general case ($d \geq 2$), if $|X|$ is not too small (when
d and t are fixed), then J is "usually" of the maximal possible rank.
But I am not yet successful in showing this, and I am not sure to what
extent this is true.

We conclude this section by giving several open problems.
First some obvious problems.

<u>Problem 4.1.</u> Prove or disprove Conjecture 1.

<u>Problem 4.2.</u> Prove or disprove Conjecture 2.

<u>Problem 4.3.</u> Classify all rigid t-designs in S^d. (The
answer would be very difficult unless Conjectures 1 and 2 were proved to
be true, although the classification still might be possible from a
logical viewpoint even if Conjectures 1 and/or 2 were false).

<u>Problem 4.4.</u> Do the above three problems for rigid t-designs
in compact symmetric spaces of rank 1. (The concept of rigid t-design
in a compact symmetric space is naturally and straightforwardly defined,
and I believe that Conjectures 1 and 2 are also true there.)

Now we mention more specific and more technical problems
concerning Conjectures 1 and 2.

<u>Problem 4.5.</u> Collect examples of rigid t-designs as many as
possible. (We know very few examples of rigid t-designs beyond tight
t-designs. It might even be true that there are no rigit t-designs in
S^d if both d and t are large.

(1) Can one show that if X is a t-design with $|X|$ close
to the lower bound (of tight t-design) then X is rigid?

(2) Can one determine when X is rigid or not for a given
X? The interesting cases are the ones where X is
obtained as an orbit of a finite subgroup G of $0(d+1)$
cf. [1, 2, 24]. It is easy to see that if $X = \vec{x}^G =$
$\{g\vec{x} \mid g \in G\} \subset S^d$ is a rigid m_2-design for a real
reflection group G and $\vec{x} \in S^d$, then $\vec{x}$ must be one
of the $d+1$ corner edges of a Weyl chamber.

<u>Problem 4.6.</u> (This is nothing but special cases of previous
Problems 4.1, 4.2, and 4.3.) Study the cases:

(1) $d = 2$ (for arbitrary t).

(2) $t = 2, 3, \ldots$ (for arbitrary d) ($t = 1$ is easily
       settled).

(3)   The simplest nontrivial open case is $(d,t) = (2,2)$.

Problem 4.7. Find further methods for the proof of Conjecture 1. The matrix J we considered before has entries not
necessarily polynomials in $\xi_1, \ldots, \xi_d$. If we take the following
$(h_1 + h_2 + \ldots + h_t + |X|) \times |X|(d+1)$ matrix J' (given below)
instead, the entries of J' are polynomials in the x_{jk} ($1 \leq j \leq |X|$,
$1 \leq k \leq d+1$),

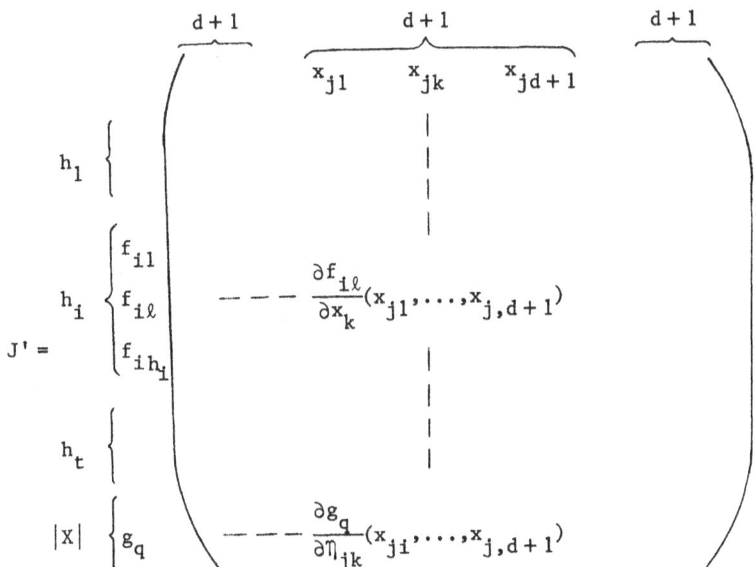

where $g_q = \eta_{q1}^2 + \ldots + \eta_{q,d+1}^2$ (of independent variables η_{qk}).
        If J' is of the maximal possible rank $(h_1 + h_2 + \ldots + h_t + |X|)$
and $|X|(d+1) - \dim O(d+1) > (h_1 + h_2 + \ldots + h_t + |X|)$, then it proves
that X is <u>not</u> rigid. (Is this J' easier to handle than J?) In a
recent letter [38], Tom Zaslavsky gave me several ideas to attack
Conjecture 1. [38] contains the following remarks:

        (1)   J being of not the maximal possible rank implies that
there exists a certain polynomial f in $x_1, x_2, \ldots, x_{d+1}$ such that
$\dfrac{\partial f}{\partial x_i} = 0$ for any $i = 1, 2, \ldots, d$ at any $(x_{j1}, x_{j2}, \ldots, x_{d+1})$ with

$1 \leq j \leq |X|$. (Note that $\dfrac{\partial f}{\partial x_i}$ are not polynomials in $x_1,\ldots,x_d$ in general, because f is a polynomial in $x_1,x_2,\ldots,x_{d+1}$.)

(2) This implies in turn that all the $(x_{j1},\ldots,x_{jd})$ $(1 \leq j \leq |X|)$ must be in the zeros of a certain nontrivial polynomial of degree at most $2t$ in $x_1,\ldots,x_d$.

(3) In particular, if $d = 1$ and if J' is not of the maximal possible rank, then $|X| \leq 2t$. This gives an alternative proof of Theorem 4.1.

REFERENCE

1. E. Bannai, Spherical t-designs which are orbits of finite groups, J. Math. Soc. Japan 36 (1984), 341-354.

2. E. Bannai, Spherical designs and group representations, Contemporary Mathematics (AMS), 34 (1984), 95-107.

3. E. Bannai, Rigid spherical t-designs in S^1 and a theorem of Y. Hong, (preprint).

4. E. Bannai, E. Bannai and D. Stanton, An upper bound for the cardinality of an s-distance subsets in real Euclidean space II, Combinatorica (Hungary) 3 (1983), 147-152.

5. E. Bannai, A. Blokhuis, P. Delsarte and J. J. Seidel, An addition formula for hyperbolic space, J. of Combinatorial Theory (A), 36 (1984), 332-341.

6. E. Bannai and R. M. Damerell, Tight spherical designs, I, J. Math. Soc. Japan, 31 (1979), 199-207.

7. E. Bannai and R. M. Damerell, Tight spherical designs, II, J. London Math. Soc. 21 (1980), 13-30.

8. E. Bannai and S. G. Hoggar, On tight t-designs in compact symmetric spaces of rank one, Proc. Japan Acad. 61A (1985), 78-82.

9. E. Bannai and S. G. Hoggar, Tight t-design in projective spaces and Newton polygons, Ars Combinatoria 20A (1985), 43-49.

10. E. Bannai and S. G. Hoggar, Tight t-designs and square-free integers, (preprint, submitted to Europ. J. Comb.).

11. E. Bannai and T. Ito, Algebraic Combinatorics I, Benjamin/Cummings Lecture Note Series in Math., Menlo Park, California, 1984.

12. E. Bannai and T. Ito, Algebraic Combinatorics, II, in preparation.

13. E. Bannai and N. J. A. Sloane, Uniqueness of certain spherical codes, Canad. J. Math. 33 (1981), 437-449.

14. A. Blokhuis, Few distance sets, Ph.D. thesis, Eindhoven, 1983.

15. I. Chavel, Eigenvalues in Riemannian Geometry, Academic Press, New York, 1984.

16. P. Delsarte, An algebraic approach to the association schemes of coding theory, Philips Research Reports Suppl. No. 10, 1973.

17. P. Delsarte, G.-M. Goethals and J. J. Seidel, Bounds for systems of lines and Jacobi polynomials, Philips Research Reports 30 (1975), Boukamp Volume 91* - 105*.

18. P. Delsarte, G.-M. Goethals and J. J. Seidel, Spherical codes and designs, Geom. Dedicata, 6 (1977), 363-388.

19. M. Deza and P. Frankl, Bounds on the maximum number of vectors with given scalar products, Proc. Amer. Math. Soc. 95 (1985), 323-329.

20. C. F. Dunkl, Discrete quadrature and bounds on t-design, Mich. Math. J. 26 (1979), 81-102.

21. A. Erdelyi et. al., Higher Transcendental Functions (Bateman Manuscript Project), McGraw-Hill, 1953.

22. P. Erdős, On a diophantine equation, J. London Math. Soc. 26 (1951), 176-178.

23. R. Gangolli, Positive definite kernels on homogeneous spaces and certain stochastic processes related to Levy's Brownian motion of several parameters, Ann. Inst. Henri Poincare 3 (1967), 121-225.

24. J.-M. Goethals and J. J. Seidel, Cubature formulae, polytopes and spherical designs, in the Geometric Vein, Springer-Verlag, 1982, 203-218.

25. S. Helgason, Differential Geometry and Symmetric Spaces, Academic Press, 1962.

26. S. G. Hoggar, t-designs in projective spaces, Europ. J. Comb. 3 (1982), 233-254.

27. S. G. Hoggar, Tight t-designs and octonions, Coxter Festschrift, Teil III, University of Giessen (1984), 1-16.

28. S. G. Hoggar, Parameters of t-designs in FP^{d-1}, Europ. J. Comb. 5 (1984), 29-36.

29. Y. Hong, On spherical t-designs in R^2, Europ. J. Comb. 3 (1982), 255-258.

30. T. Koornwinder, The addition formula for Jacobi polynomials and spherical harmonics, SIAM J. Appl. Math. 25 (1973), 236-246.

31. D. G. Larman, C. A. Rogers and J. J. Seidel, On two distance sets in Euclidean spaces, Bull. London Math. Soc. 9 (1977), 261-267.

32. H. Morikawa, Some results on harmonic analysis on compact quotients of Heisenberg groups, Nagoya Math. J. 99 (1985), 45-62.

33. A. Neumaier, Combinatorial configurations in terms of distances, T. H. E. (Eindhoven) Memorandum 81-90, 1981.

34. P. Seymour and T. Zaslavsky, Averaging set: A generalization of mean values and spherical designs, Advances in Math. 52 (1984), 213-240.

35. G. Szegö, Orthogonal Polynomials, 4th edition, Amer. Math. Soc. Providence, R.I., 1975.

36. P. Terwilliger, A characterization of P- and Q-polynomial association schemes, (preprint).

37. J. A. Wolf, Spaces of Constant Curvature, McGraw-Hill, 1967.

38. T. Zaslavsky, Personal communication.

METRIC AND GEOMETRIC PROPERTIES OF SETS OF PERMUTATIONS

P.J. Cameron

School of Mathematical Sciences
Queen Mary College
Mile End Road
London
E1 4NS

ABSTRACT

The purpose of this paper is to survey some recent results on sets of permutations. These results involve a nice fusion of the two themes of this conference, algebraic and extremal methods. Many of these results were obtained in collaboration with M. Deza and P. Frankl, and can be found in our joint paper (to appear), or in the references therein; so the organising committee of the conference must bear dual responsibility!

INTRODUCTION

I will be discussing a translation, to families of permutations, of problems which have been well studied in the context of families of sets. In this context, typical *hypotheses* on a family F include:

(a) Conditions on cardinalities of intersections of the sets (for example, prescribing the numbers of cardinalities of pairwise intersections). If all the sets have the same size, this is equivalent to conditions on the Hamming distances of the corresponding zero-one vectors.

(b) Design conditions (typically: any set of t points is contained in a constant number of members of F).

(c) Closure conditions, such as closure under symmetric difference (linear codes) or under supersets (filters).

Typical *questions* are

(i)   bounds on $|F|$ (we expect, for example, upper bounds in case (a), or lower bounds in case (b));

(ii)   characterisation in extremal cases.

In order to interpret these hypotheses for permutations, it is helpful to represent a permutation by its graph That is, take $N = \{1, 2, ..., n\}$; a permutation g of N is represented by the n-subset $\Gamma(g) = \{(i, g(i)) | i \in N\}$ of the n^2-set $N \times N$. A subset of $N \times N$ is the graph of a permutation if and only if it is a transversal for both rows and columns of $N \times N$. We have

$$|\Gamma(g_1) \cap \Gamma(g_2)| = |\{i | g_1(i) = g_2(i)\}|$$

$$= |\text{fix}(g_1{}^{-1}g_2)|,$$

where $\text{fix}(g) = \{i | g(i) = i\}$. So the analogue of a metric hypothesis of type (a) is a restriction on the values of $|\text{fix}(g_1^{-1}g_2)|$ for $g_1, g_2 \in F$, $g_1 \neq g_2$.

Now consider hypothesis (b). A t-element subset of $N \times N$ cannot be contained in the graph of a permutation unless it is the graph of a *partial permutation*, a bijection between t-subsets of N. So the appropriate t-design condition for F is the following; given any partial permutation f of N of cardinality t, a constant number λ of members of F contain f. We say that F is *uniformly t-transitive* if this holds; *sharply t-transitive* if it holds with $\lambda = 1$.

For $t = 1$, we say simply *uniformly* (or *sharply*) *transitive*. The following simple extremal result characterises uniformly transitive sets. The proof is a simple application of Cauchy's inequality, and the proof generalises a result of Cauchy and Frobenius for groups.

Theorem 1. Let F be any set of permutations. Then the average value of $|\text{fix}(g_1{}^{-1}g_2)|$, for $g_1, g_2 \in F$, is at least 1, with equality if and only if F is uniformly transitive.

T. Ito, in his talk at the meeting discussed the connection between uniform transitivity and Delsarte's concept of T-designs in

association schemes. Subsequently, he made the following observation.
Let B be the set of bocks of a t-design (in the usual sense) on
$\{1, \ldots, n\}$ with block size k, and let

$$F = \{g \in S_n | g(\{1, \ldots, k\}) \in B\}$$

Then F is uniformly t-transitive.

The obvious closure condition for permutations, and the
only one I consider, is that F be closed under composition, that is, F
is a group of permutations. This simplifies assumptions (a) and (b) as
follows. If F is a group, then

(i) $\{|\text{fix}(g_1{}^{-1}g_2)| \,| g_1, g_2 \in F, g_1 \neq g_2\}$
$= \{|\text{fix}(g)| \,| g \in F, g \neq 1\}.$

(ii)   F is uniformly t-transitive if and only if it is
t-transitive, that is, every partial permutation of
size t is contained in some member of F.

(The first fact, which is trivial, can be compared to the relation
between distances and weights in linear codes. The second holds
because, in a t-transitive group, the set of elements containing a
given partial permutation of size t is a coset of a t-point stabiliser,
and all such stabilisers are conjugate).

Sharply t-transitive sets correspond to well-known
geometric objects (in particular, Latin squares for t = 1, projective
planes for t = 2); their determination for $t \geqslant 2$ is highly non-trivial,
as C. Lam's talk at the conference bears out. On the other hand, all
sharply t-transitive groups for $t \geqslant 2$ have been known since 1930,
when Zassenhaus completed the cases t = 2, 3; the case $t \geqslant 4$ had been
done much earlier by Jordan. (See Passman (1968) for an account of
this work). Moreover, the classification of finite simple groups leads
to a determination of all t-transitive of groups for $t \geqslant 2$.

2 Bounds for s-distance sets and groups

Let F be a family of permutations of N = $\{1, \ldots, n\}$, and set

$$L = \{|\mathrm{fix}\,(g_1^{-1}\,g_2)|\,|g_1,\ g_2 \in F,\ g_1 \neq g_2\},$$

and

$$s = |L|. \quad \text{Given n and s, how large can } |F| \text{ be?}$$

For $s = 1$, the question has been considered by many authors, and the results are summarised in Deza (1983). Such sets of cardinality cn^2, where c is about ¼, have been constructed by Heinrich, van Rees and Wallis (1977). N. Alon pointed out at the meeting that, using this result, a general lower bound can be found as follows. Take a one-distance set E of permutations of an (n/s)-set, with $|E| = c(n/s)^2$, and let F be the "direct sum" of s copies of E (that is, the set of s-tuples of elements of E, acting on pairwise disjoint sets). If $|\mathrm{fix}(g_1^{-1}g_2)| = \ell$ for $g_1,\ g_2 \in \overline{c},\ g_1 \neq g_2$, then

$$\mathrm{fix}(g_1^{-1}g_2) \; = \; in/s + (s-i)$$

for $g_1,\ g_2 \in F,\ g_1 \neq g_2$, where i is the number of coordinates in which the s-tuples g_1 and g_2 agree $(i - 0,\ 1,\ ...,\ s-1)$; so F is an s-distance set of permutations of an n-set, and

$$|F| \; = \; (c^s/s^{2s})n^{2s}.$$

Regarding permutations as n-subsets of an n^2-set, we see from the theorem of Ray-Chaudhuri and Wilson (1975) that an s-distance set F satisfies

$$|F| \; \leqslant \; \begin{bmatrix} n^2 \\ s \end{bmatrix} \sim (1/s!)n^{2s}.$$

The lower and upper bounds have the same dependence on n, differing by a ratio which is a function of s. Using a little character theory of the symmetric group, it is possible to reduce this ratio from factorial to exponential in s.

Theorem 2. Let F be an s-distance family of permutations of an n-set.
Then

$$|F| \;<\; \sum_{\substack{\chi \,\in\, \mathrm{Irr}(Sn) \\ \dim(\chi) \,<\, s}} \chi(1)^2$$

Remark.   There is a well-known bijection between irreducible characters of the summetric group S_n and partitions of n.   The dimension of a character is n minus the largest part of the corresponding partition.   For fixed s and $n > 2s$, there are p(s) characters of dimension s, where p is the partition function; the degree of any such character is a polynomial in n of degree s, with leading coefficient (1/s!).   So, for fixed s and large n, the order of magnitude of the bound in Theorem 2 is $(p(s)/(s!)^2)n^{2n}$.

Outline of proof.   Let π be the permutation character of S_n (so that $\pi(g) = |\mathrm{fix}(g)|$), and let $f_L(x)$ be the polynomial.

$$\prod_{\ell \,\in\, L} (x - \ell).$$

Then, for $g_1, g_2 \in F$, we have

$$f_L\left(\pi(g_1{}^{-1}g_2)\right) \;=\; \begin{cases} f_L(n) & \text{if } g_1 = g_2 \\ 0 & \text{if } g_1 \neq g_2 \end{cases} \qquad \ldots (1)$$

Now $f_L(\pi)$ is a generalised character of S_n, that is, the difference between two ordinary characters χ_1 and χ_2. Let M_1 and M_2 be matrix representations affording χ_1 and χ_2.   Then (1) asserts that the matrices

$$\begin{bmatrix} M_1(g) & 0 \\ 0 & M_2(g) \end{bmatrix} \qquad \ldots (2)$$

for $g \in F$ form an orthogonal set with respect to the hyperbolic inner product

$$\left[\begin{bmatrix} A_1 & 0 \\ 0 & A_2 \end{bmatrix}, \begin{bmatrix} B_1 & 0 \\ 0 & B_2 \end{bmatrix} \right] = \mathrm{trace}\,(A_1{}^* B_1) - \mathrm{trace}\,(A_2{}^* B_2).$$

Hence $|F|$ does not exceed the dimension of the space spanned by the matrices (2) for all $g \in S_n$, which is the sum of squares of the degrees of the irreducible characters appearing in X_1 or X_2.

Since $X_1 - X_2$ is a polynomial of degree s in π, the irreducible characters have dimension at most s, by a result of Frobenius.

Problem 1. Narrow the gap between the upper and lower bounds!

In the case where F is a group, a much better result holds.

Theorem 3. Let F be a group of permutations of an n-set, and let $L = \{\pi(g) \mid g \in F, g \neq 1\}$. Then $|F|$ divides

$$f_L(n) \;=\; \prod_{\ell \in L} (n-\ell).$$

At the conference, I attributed this result to Kiyota (1979), but M. Deza subsequently drew my attention to a proof contained in a paper of Blichfeldt (1904). In fact, Blichfeldt attributes the theorem to Maillet (1895), but the theorem stated and proved by Maillet is quite different.

Proof.

$$f_L\,(\pi(g)) \;=\; \begin{cases} f_L(n) & \text{if } g = 1 \\ 0 & \text{if } g \neq 1 \end{cases},$$

so $f_L(\pi)$ is a multiple of the regular character ρ of F, given by $\rho(g) = |F|$ if $g = 1$, 0 if $g \neq 1$. The result follows.

Note that this bound is at most n^s, where $s = |L|$.

A group of permutations attaining this bound is called *sharp*. Such groups are studied in Ito and Kiyota (1981) and Cameron, Deza and Frankl (to appear).

There are a number of specific situations where the quantity $f_L(n)$ occurs, either as a bound or as an exact value of $|F|$.

For example:

Theorem 4.   Let F be a set of permutations with
$|\text{fix}(g_1^{-1} g_2| < s$ for g_1, $g_2 \in F$, $g_1 \neq g_2$, and set $L = \{0, ..., s-1\}$.
Then $|F| \leqslant f_L(n) = n(n-1) ... (n - s + 1)$, with equality if and only if
F is sharply s-transitive.

Proof.   Given s a distinct points i_1, ..., i_s, the s-tuples
$(g(i_1), ..., g(i_s))$, for $g \in F$, are all distinct (else two members of F
intersect in at least s points). So $|F|$ does not exceed the number of
s-tuples of distinct points, which is $f_L(n)$. The characterisation of
equality is clear.

The geometric sets of the next section are generalisations
of sharply s-transitive sets.

It is appropriate to mention here a result of Deza and
Frankl (1977). A set F of permutations is said to be of *type* (L, n) if
$\text{fix}(g_1^{-1}g_2) \in L$ for all g_1, $g_2 \in F$, $g_1 \neq g_2$.

Theorem 5.   (i)   If F and F´ are sets of type (L, n) and
(L´, n) respectively, where $L´ = \{0, ..., n-1\} \backslash L$, then

$$|F| \cdot |F´| \leqslant n!$$

(ii)   If there exists a set of type (L´, n) and cardinality
$f_{L´}(n)$, where $L´ = \{0, ..., n-1\} \backslash L$, then any set of type (L, n) has
cardinality at most $f_L(n)$.

Problem 2.   It follows from Theorem 5 that a set of type
$(\{t, t+1, ..., n-1\}, n)$ has cardinality at most $(n-t)!$, provided that a
sharply t-transitive set exists. Show that the assumption can be
replaced by the condition $n \geqslant n_0(t)$. The same condition should also
ensure that the only extremal sets are the cosets of t-point
stabilisers. (This would be the analogue of the Erdös-Ko-Rado theorem
(1961)).

Problem 3.   Since sharply t-transitive sets are so rare, we

could ask whether, for fixed t, there is always a set of type
$(\{0, ..., t-1\}, n)$ and cardinality $(1 - o(1))n^t$. (This would be the
analogue of a recent theorem of Rödl (1985) for sets).

3 Geometric sets and groups.

A permutation geometry is an analogue, for permutations, of
a matroid.

A matroid can be specified by its family of flats. The
non-trivial part of the definition asserts that, if F is a flat and x a
point outside F, there is a unique flat F′ containing F $\cup$ {x} with rank
(F′) = rank(F) + 1. (The rank of a flat F is the length of the longest
chain of flats with greatest element F).

Following our observation that a subset of a permutation
must be a partial permutation, we are led to define a *permutation
geometry* as a family G of partial permutations, closed under
intersection, such that if F $\in$ G and (x, y) $\in$ N × N are such that
F $\cup$ {(x, y)} is a partial permutation, then there exists a unique F′ $\in$ G
containing F $\cup$ {(x, y)} such that rank(F′) = rank(F) + 1.

This definition ensures that the hyperplanes (maximal flats)
are permutations. A set of permutations is *geometric* if it is the
hyperplane family of a permutation geometry. The definition can be
made intrinsic: F is geometric if the meet semilattice it generates (its
closure under intersection) is a permutation geometry. The definition
can be further modified so that permutation geometries are not
mentioned.

For example, let n = 4; let G consist of the empty set, all
partial permutations of cardinality 1, and all even permutations of N.
Then G is a permutation geometry, and so the alternating group A_4 is
a geomtric set. A similar construction shows that any sharply
t-transitive set of permutations is geometric.

We will always make the further assumption that a flat of
rank i has cardinality ℓ_i for i = 0, ..., s (where s is the rank of a
maximal flat, so that $\ell_s = n$), and set L = $\{\ell_0, ..., \ell_{s-1}\}$. We call
(L, n) the *type* of the geometry, or the geometric set. Thus A_4 has
type $(\{0, 1\}, 4)$. (This condition parallels the definition of a perfect
matroid design). Since the geometry is intersection-closed, we have

$$|\mathrm{fix}(g_1^{\,-1}g_2)| \in L$$

for all g_1, g_2, $\in L$, $g_1 \neq g_2$; so this is compatible with our earlier definition of type. Also, a straightforward argument shows that

$$|F| = f_L(n) = \prod_{\ell \in L} (n-\ell).$$

We give one more example, to illustrate a technique called blowing-up. From A_4, we wish to construct a geometric set of type $(\{0, m\}, 4m)$ by replacing each point with an m-set. For each $m \times m$ cell of the $4m \times 4m$ set $N \times N$, we require a partition into partial permutations of size m; this is just a Latin square of order m. Numbering these permutations from 1 to m, we see that, for each even permutation of the four cells, we need a set A of m^2 4-tuples over the alphabet $\{1, ..., m\}$ with the property that, given any two coordinates and any two letters of the alphabet, a unique member of A has those letters in those coordinate positions. (The set A describes the choice of partial permutations in each cell). Such an orthogonal array A is equivalent to a pair of orthogonal Latin squares of order m, and exists for all m except $m = 2$ and $m = 6$. (There are geometric sets of type $(\{0, 2\}, 8)$ and $(\{0, 6\}, 24\}$ which are not obtained by blowing up A_4).

The blowing up process can be formulated in general for any geometric set; the role of A is played by an object called a *transversal matroid design*. See Cameron et al. (to appear) for details.

A *geometric group* is a geometric set which is a group. Note that any geometric group attains the bound in Theorem 3, and so is sharp. (Easy exanples show that the converse is false). The structure of geometric groups is so much tighter that a compelete determination seems possible; work in progress by T. Maund is directed towards this goal, and the rank 2 case has been completed (Maund (1986)). The following remarks will indicate some of the flavour of this work.

Blowing-up has a nice algebraic description for groups. If $1 < s < n$, a blow-up of F is an extension of an abelian group A by F, satisfying some extra conditions. These conditions only involve the action of F on A, not the splitting or non-splitting of the extension.

So the determination falls into two parts: finding all *deflated* groups (those which are not blow-ups of smaller groups), and then blowing them up in all possible ways.

It is natural to use induction, and so to consider first the case s = 2. In this case, the group F permutes 2-transitively the set of domains of rank 1 flats of the geometry; so we examine the list of 2-transitive groups and decide whether each can be a quotient of a geometric group. The second question involves analysis of the representation theory of the deflated group, since we can reduce to the case where A is elementary abelian, that is, a GF(p)F-module for some prime p. In this way, Maund proved the following theorem:

Theorem 6. A geometric group of type ({0, m}, nm) is one of the following:

(i)   A blow-up of S_2 or S_3 (n = 2 or 3, m arbitrary);

(ii)   a blow-up of AGL(1, q) (n = q, m = q^d);

(iii)   a blow-up of GL(2, q) (n = q+1, m = $(q-1)q^d$);

(iv)   sharply 2-transitive (n a prime power, m = 1);

(v) $Z_{\frac{1}{2}(q-1)} \times$ PSL(2, q), q $\equiv$ 3 (mod 4) (n = q+1, m = $\frac{1}{2}(q-1)$);

(vi) $Z_{q-1} \times$ Sz(q) (n = q^2+1, m = q-1, q an odd power of 2);

(vii)   PGL(3, q) (n = q^2+q+1, m = q(q-1), q = 2 or 3).

The abelian groups used for the blow-ups in (ii) and (iii) are vector spaces over GF(q).

4 *Some generalisations*

Suppose that F is a geometric group, and G the corresponding permutation geometry. Consider the set M of all

domains of partial permutations in G.   It is straightforward to show
that

    (i)   M is the set of flats of a matroid;

    (ii)   G consists of all restrictions of members of F
           to members of M;

    (iii)   F is a group of automorphisms of M, which acts
           sharply transitively on the set of bases of the
           matroid.

So a generalisation of the problem of finding geometric
groups is that of finding sharply basis-transitive groups (or sets!) of
automorphisms of matroids.   For simple matroids, all basis-transitive
examples were found by Kantor (1985), with the help of the
classification of finite simple groups.   In most cases, the existence of
sharply basis-transitive subgroups is easy to decide.   However, for
non-simple matroids, the problem is harder.   In the rank 2 case, the
groups are just those groups of automorphisms of complete
multipartite digraphs which act sharply transitively on edges; these
are studied in Babi et al. (1981) and Cameron (to appear).

Another possible generalisation is described by Deza and
Laurent (to appear) (see also Cameron et al. (in preparation)).   Noting
that permutations of N can be identified with subsets of N × N, we can
pose similar problems for subsets of N^d which are transversals for all
the coordinate hyperplanes, for any d $\geqslant$ 2.   If one coordinate (say the
first) is distinguished, such a subset F has the form

$$F \; = \; \{(i, \; g_1(i),..., \; g_{d-1}(i)) \,|\, i \in N\}.$$

where $g_1, \; ..., \; g_{d-1}$ are permutations, and so can be identified with the
element $(g_1, \; ..., \; g_{d-1})$ of S_n^{d-1}.   The cardinality of the intersection of
two such subsets F and F′ is equal to

$$\left| \bigcap_{i=1}^{d-1} \; \text{fix}(g_1{}^{-1}g'_i) \right|.$$

So we can ask for bounds on $|F|$ when the cardinalities of these intersections lie in a prescribed set L.   The analogue of a geometric set of permutations can be defined; the role of permutation geometries is played by the so-called *injection geometries*.   We can also specialise the questions to the case when F corresponds to a subgroup of S_n^{d-1}.

Problem 4.   Prove that, if F is a subgroup of S_n^{d-1} such that

$$L = \left\{ \left| \bigcap^{d-1} \mathrm{fix}(g_i^{-1}g'_i) \right| \middle| g, g' \in F, g \neq g' \right\},$$

then $|F| \leqslant f_L(n)^{d-1}$, with equality if and only if F is geometric.

These concepts are less amenable to algebraic techniques than those of Section 1.   For example,

(i)   a set F of diagonals of N^d may yield a group when one coordinate is distinguished, but not when another is distinguished;

(ii)   conjugate subgroups of S_n^{d-1} may have quite different types (the *type* being the set L above).

5 An application

Delsarte (1978) proved the following theorem:

Theorem 7.   Let t, n, m be given with $t \leqslant n \leqslant m$.   Let V and W be vector spaces over GF(q) with $\dim(V) = n$, $\dim(W) = m$.   Then there is a set F of linear transformations from V to W with the following property: given any linear transformation T from a t-dimensional subsapce of V to W, there is a unique member of F which extends T.   Moreover, F can be chosen to be a GF(q)-subspace of Hom(V, W).

Such sets F are called *Singleton systems*.   They can be considered as a vector space analogue of orthogonal arrays.   We could try to manufacture the vector space analogue of sharply t-transitive sets by taking V = W and requiring all the linear transformations to have kernel zero. Thus, we define a *vector transversal design* of type (t, n) over GF(q) to be a family F of nonsingular linear transformations form V to V, where V is n-dimensional over GF(q), such that, given any one-to-one linear transformation T from a t-dimensional subspace of v into V, there is a unique member of F which extends T.

Problem 5. for which t, n, q do vector transversal designs exist?

It is readily checked that a vector transversal design is precisely a geometric subset of GL(n, q) of type ($\{1, q, ..., q^{t-1}\}, q^n$) (acting on V). Using lemmas of O'Nan (1985) and Cameron and Taylor (to appear), the following can be shown:

Theorem 8. If a vector transversal design of type (t, n) exists, then either t = 1, or

$$n \ \leqslant \ \begin{cases} \tfrac{1}{2}(5t+1) & t \text{ odd} \\ \tfrac{1}{2}(5t-2) & t \text{ even} \end{cases}$$

See Cameron (to appear) for the proof.

Examples exist with t = n (the whole of GL(n, q), and with t = 1 (identifying V with GF(q^n), take for F the set of multiplications by nonzero elements of GF(q^n)). Apart from these, just one example is known. The group GL(4, 2) is isomorphic to A_8; its subgroup A_7 is a vector transversal design of type (3, 4) over GF(2). Note that this attains the bound in Theorem 8.

REFERENCES

Babai, L., Cameron, P.J., Deza, M. & Singhi, N.M. (1981). On sharply
          edge-transitive permutation groups. J. Algebra 73, 573-585.

Blichfeldt, H.F. (1904). A theorem concerning the invariants of
          linear homogeneous groups, with some applications to
          substitution groups. Trans. Amer. Math. Soc. 5, 461-466.

Cameron, P.J. (to appear). Digraphs admitting sharply edge-transitive
          automorphism groups. Europ. J. Combinatorics.

Cameron, P.J. (to appear). Geometric sets of permutations. Geometriae
          Dedicata.

Cameron, P.J., & Deza, M. (1979). On permutation geometries. J.
          London Math. Soc. (2), 20, 373-386.

Cameron, P.J., Deza, M. & Frankl, P. (to appear). Sharp sets of
          permutations. J. Algebra.

Cameron, P.J., Deza, M., & Frankl, P. (in preparation). Intersection
          theorems for permutation groups.

Cameron, P.J., & Taylor, D.E. (to appear). Stirling numbers and
          affine equivalence. Utilitas Math.

Delsarte, P. (1978). Bilinear forms over a finite field, with
          applications to coding theory. J. Combinatorial Theory (A),
          25, 226-241.

Deza, M. (1983). Some new generalisations of sharply t-transitive
          groups and sets. In Combinatorics '81 (Rome 1981),
          pp. 195-314. Amsterdam: North Holland.

Deza, M., & Frankl, P. (1977). On the maximum number of permutations
          with given maximal or maximal distance, J. Combinatorial
          Theory (A), 22, 352-360.

Deza, M., & Laurent, M. (to appear). Squashed geometries.

Erdös, P., Ko, C., & Rado, R. (1961). Intersection theorems for
          systems of finite sets, Quart. J. Math. Oxford (2), 12,
          313-320.

Heinrich, K., Van Rees, G.H.J., & Wallis, W.D. (1979). A general
          construction for equidistant permutation arrays. In Graph
          Theory and Related Topics (Waterloo 1977), pp. 247-252.
          New York: Academic Press.

Ito, T., & Kiyota, M. (1981). Sharp permutation groups, J. Math. Soc.
          Japan 33, 435-444.

Kantor, W.M. (1985). Homogeneous designs and geometric lattices. J.
          Combinatorial Theory (A), 38, 66-74.

Kiyota, M. (1979). An inequality for finite permutation groups. J. Combinatorial Theory (A), 27, 119.

Maillet, E. (1895). Sur quelques propriétés des groupes des substitutions d'order donné. Ann. Fac. Sci. Toulouse, 1-22.

Maund, T. (1896). Geometric groups. Dissertation, Oxford University.

O'Nan, M.E.   (1985). Sharply 2–transitive sets of permutations. In Proceedings of the Rutgers Group Theory Year 1983–1984. (Ed. M. Aschbacher et al). pp. 63–67. Cambridge: Cambridge Univ. Press.

Passman, D.S. (1968). Permutation Groups. New York: Benjamin.

Ray–Chaudhuri, D.K., & Wilson, R. (1975). On t–designs, Osaka J. Math. 12, 737–744.

Rödl, V. (1985). On a packing and covering problem. Europ. J. Combinatorics. 6, 69–78.

INFINITE GEOMETRIC GROUPS AND SETS

P.J. Cameron
Queen Mary College, London, U.K.

M. Deza
CNRS, Paris, France.

N.M. Singhi
Mehta Institute of Fundamental Research, Allahabad, India.

ABSTRACT

We investigate geometric groups and sets of permutations
of an infinite set. (These are a generalisation of sharply
t-transitive groups and sets). We prove non-existence of
groups, and give constructions of sets, for certain
parameters. This work was done while the authors were
visiting the Ohio State University, to whom we express our
gratitude.

1 INTRODUCTION

It is known that sharply t-transitive groups of permutations
of an infinite set exist only for $t \leqslant 3$ (Tits (1952)), while sharply
t-transitive sets exist for all t (Barlotti & Strambach 1984).

Geometric groups and sets of permutations have been proposed
as a natural generalisation of sharply t-transitive groups and sets
(Cameron & Deza (1979)). Our purpose is to investigate such objects on
infinite sets. Not surprisingly, we give nonexistence results for
groups, and constructions for sets.

Let $L = \{\ell_0, \ell_1, \ldots, \ell_{s-1}\}$ be a finite set of natural
numbers, with $\ell_0 < \ldots < \ell_{s-1}$. The permutation group G on the set X is
a *geometric group of type* L if there exist points $x_1, \ldots, x_s \in X$ such
that

(i)   the stabiliser of $x_1, \ldots, x_s$ is the identity;

(ii)   for $i < s$, the stabiliser of $x_1, \ldots, x_i$ fixes ℓ_i
       points and acts transitively on its non-fixed points.

Theorem 1. There is no infinite geometric group of type
$\{0, m, 2m, \ldots, (p-1)m\}$, where p is prime and p divides m.

This theorem is proved in Section 2.   The case $p = m = 2$ is an infinite analogue of a theorem of Tsuzuku (1968).   The theorem is new only for $p < 3$, since the non-existence of infinite geometric groups of type $\{0, m, ..., (t-1)m\}$ for any $t \geqslant 4$ follows from a theorem of Yoshizawa (1979) (see Cameron *et al.* (to appear)).   However, the case $p = 2$ is the most interesting.   It has the following consequence, also derived in Section 2:

Corollary 2.   Let $L = \{\ell_0, ..., \ell_{s-1}\}$, with $\ell_0 < ... < \ell_{s-1}$.   If $\ell_{i+1} - \ell_i$ is even for some $i \in \{0, ..., s - 2\}$, then no geometric group of type L exists.

Geometric sets are harder to define; but our examples (in common with all known finite examples) satisfy an additional, simplifying condition.   Let $L = \{\ell_0, ..., \ell_{s-1}\}$ with $\ell_0 < ... < \ell_{s-1}$.   The set S of permutations of X is a *special geometric set of type* L if there is a matroid M of rank s on X satisfying

     (i)   any element of S is an automorphism of M;

     (ii)   for $i < s$, a flat of M of rank i has cardinality ℓ_i;

     (iii) for any $g, h \in S$, $\{x \in X | g(x) = h(x)\}$ is a flat of M;

     (iv)   S is sharply basis-transitive, that is, if
           $(x_1, ..., x_s)$ and $(y_1, ..., y_s)$ are bases of M, there
           is a unique $g \in S$ with $g(x_i) = y_i$ for $i = 1, ..., s$.

It is easily seen that any geometric group is a special geometric set; the bases of M are the s-tuples with the property of the definition.   A geometric set in the sense of Cameron and Deza (1979) is special if and only if it is unisupported (all the matroids $\hat{g}$ coincide) and consists of automorphisms of this common matroid .

In Section 3, we prove:

Theorem 3.   There exist special geometric sets of permutations of a countable set, of each of the following types:

(i) $L = \{0, m, 2m, ..., (t-1)m\}$ for any t, $m > 1$;

(ii) $L = \{0, 1, q, ..., q^{t-1}\}$ for any $t > 1$ and any prime
power q.

2   Geometric Groups

We begin with a lemma. This lemma, in more general form,
is due to Yoshizawa (1979); we repeat the proof for completeness.
However, by Yoshizawa's theorem, the lemma is vacuous unless $p < 3$.

Lemma 2.1. Let p be a prime. Suppose that G is p-transitive on an
infinite set, and that the stabiliser of p points is finite with order
divisible by p. Then some element of order p fixes infinitely many
points.

Proof. Suppose that G is a counterexample to the lemma. Let $g \in G$
be an element of order p with (finite) fixed point set F (with $|F| > p$)
and (infinitely many) cycles C_1, C_2, ... For each $i \in \mathbb{N}$, g normalises
the pointwise stabiliser H_i of C_i; so there is an element
$h_i \in H_i$ of order p which commutes with g. Thus h_i maps F to F. But
infinitely many of the elements h_i are different, since by assumption
an element of order p can belong to only finitely many subgroups H_i.
Thus, the setwise stabiliser of F is infinite, and so is its pointwise
stabiliser, contrary to assumption.

If L is finite and $0 \in L$, a geometric group G of type L is
transitive, and the stabiliser of a point fixes ℓ_1 points, where
$\ell_1 = \min(L \setminus \{0\})$. These fixed point sets are blocks of imprimitivity for
G; for brevity, we call them blocks. Note that the stabiliser of a point
acts on the set of its non-fixed points as a geometric group H of type
$L' = \{\ell - \ell_1 | \ell \in L, \ell \neq \ell_0\}$. We require a lemma about the kernel of
the action of G on the set of blocks.

Lemma 2.2. Let G be geometric of type L, where L is finite and $0 \in L$.
Then the kernel K of the action of G on its set of blocks is
semiregular; that is, no non-identity element fixes every block and a
point in one of them.

Proof. Observe first that K is finite; indeed, $|K| \leqslant \ell_1{}^s$, where $s = |L|$, because the stabiliser of s independent points is trivial. The lemma is proved by induction on s, being clear when $s = 1$. We suppose the result true for L'-geometric groups with $|L'| = s - 1$.

Let H be the stabiliser of a point x, acting on its non-fixed points. By the induction hypothesis, the stabiliser of all H-blocks is semiregular. Each H-block is a union of G-blocks; so, if K is the kernel of the action on G-blocks, then $K_{xy} = 1$ whenever x and y lie in different G-blocks. Thus, as x runs over a set of representatives for the G-blocks, the subgroups K_x all have the same order and intersect pairwise in the identity. Because there are infinitely many G-blocks but K is finite, this requires $K_x = 1$.

Proof of Theorem 1. Suppose that G is a geometric group of type $\{0, m, ..., (p - 1)m\}$, where p is prime and $p|m$, and let K be the kernel of the action of G on blocks.

Any two elements of order p in $G \backslash K$ are conjugate. For, let g and h be two such elements; let $(x_1, ..., x_p)$ be a cycle of g with its points in distinct blocks, and $(y_1, ..., y_p)$ a cycle of h with the same property. Then $\{x_1, ..., x_p\}$ and $\{y_1, ..., y_p\}$ are bases of the matroid M. So h is the unique element of G having a cycle $(y_1, ..., y_p)$; and there exists $k \in G$ with $k(x_i) = y_i$ for $i = 1, ..., p$. Then kgk^{-1} has a cycle $(y_1, ..., y_p)$, and so $kgk^{-1} = h$.

By lemma 2.2, it follows that any two elements of order p which fix a point are conjugate.

Now G acts p-transitively on the set of blocks, and the stabiliser of p blocks is finite with order divisible by p (its order is a multiple of m^{p-1} and a divisor of m^p). Let g be an element of order p fixing $p - 1$ blocks $B_1, ..., B_{p-1}$ pointwise and another block setwise. By the above remarks and Lemma 2.1, g fixes infinitely many blocks setwise, say $B_1', B_2', ...$

For any $i \geqslant p$, g normalises the subgroup H_i fixing $B_1', B_2', ..., B_{p-1}'$ pointwise and B_1' setwise; so H_i contains an element h_i of order p commuting with g. Then h_i fixes setwise the fixed point set $B_1 \cup ... \cup B_{p-1}$ of g, and preserves the block system; so it fixes $B_1, ..., B_{p-1}$ setwise. Also, the elements h_i are all distinct, since if $h_i = h_j$ then h_i fixes $B_i', B_j', B_2', ..., B_p'$ (a total of p blocks) pointwise. Thus the setwise stabiliser of B_1, ..., B_{p-1}, $B_1{}'$ is infinite, a contradiction.

Proof of Corollary 2. Well-known necessary conditions for the existence of perfect matroid designs (see Cameron & Deza (1979)) show that $\ell_{i+1} - \ell_i$ divides $\ell_{s-1} - \ell_{s-2}$; so we may assume that $\ell_{s-1} - \ell_{s-2}$ is even. Then the stabiliser of $s-2$ independent points, acting on its set of non-fixed points, is geometric of type $\{0, \ell_{s-1} - \ell_{s-2}\}$, contradicting Theorem 1.

3　Geometric Sets

We begin with some general remarks.

1. The property of being a special geometric set of type L is unaffected by left or right multiplication by a fixed permutation. So we may assume, without loss, that a special geometric set contains the identity.

2. If $\min(L) = \ell_0 > 0$, then the existence of special geometric sets of type L and of type $L - \ell_0 = \{\ell - \ell_0 \,|\, \ell \in L\}$ are equivalent. For we may add points fixed by every permutation; conversely, if S has type L and contains the identity, then the members of S have ℓ_0 common fixed points, which may be deleted.

3. If $0 \in L$, then the existence of a special geometric set of type L implies the existence of one of type $L\setminus\{0\}$ (for example, all permutations in L which fix some given point).

As an illustration, Theorem 3(ii) implies the existence of special geometric sets of type $\{0, 1, 3, 7, ..., 2^t-1\}$.

Proof of Theorem 3. The strategy is to prescribe in advance the matroid M, and to construct freely the required set of permutations.

(i) Consider the case $L = \{0, m, ..., (t-1)m\}$. Let X be the disjoint union of countably many m-sets X_1, X_2, ..., called blocks, where $X_i = \{x_{i0}, x_{i1}, ..., x_{im-1}\}$. We construct a sequence of pairs

(G_n, m_n), where G_n is a set of bijections between subsets of X, and m_n is a positive integer, so that the following conditions are satisfied, where $A_n = X_1 \cup \ldots \cup X_{m_n}$:

(i) Each member of G_n is contained in a unique member of G_{n+1}, and $m_n < m_{n+1}$.

(ii) The domain and range of each member of G_n are unions of blocks and contain A_n.

(iii) If $(y_1, \ldots, y_t)$ and $(z_1, \ldots, z_t)$ are t-tuples of elements of A_n, the members of each tuple lying in distinct blocks, then a unique member of G_n carries y_i to z_i for $i = 1, \ldots, t$.

(iv) If $g \in G_n$ and $g(x_{ij}) = x_{k\ell}$, then $g(x_{ij+s}) = x_{k\ell+s}$ for $s = 1, \ldots, m - 1$, where the second subscript is taken mod m.

It is clear that a starting pair (G_0, m_0) satisfying (ii)–(iv) exists.

Suppose that (G_n, m_n) has been constructed. The construction of (G_{n+1}, m_{n+1}) involves two steps.

(a) First, let m_{n+1} be the largest index of a block contained in the domain or range of an element of G_n, and $A_{n+1} = X_1 \cup \ldots \cup X_{m_{n+1}}$. For each pair of t-tuples from A_{n+1} as in (iii), if there is not already a member of G_n carrying the first to the second, adjoin such a bijection between the unions of the blocks containing the two t-tuples, in such a way that (iv) is satisfied.

(b) Now extend each bijection so that its domain and range contain A_{n+1}. For example, if g is not defined on the block $B_i \in A_{n+1}$, select the first block not already used in the constructon, say B_j, and let g map B_i to B_j so that (iv) is satisfied.

It is now clear that all the conditions hold for (G_{n+1}, m_{n+1}).

Now let G be the set of permutations of X obtained as

direct limits (or unions) of sequences (g_n) of bijections, where $g_n \in G_n$ and $g_n \subseteq g_{n+1}$. (All such direct limits are permutations, by (ii)). Clearly any element of G permutes the blocks among themselves; so G consists of automorphisms of the matroid M whose i-flats are unions of i distinct blocks for $i < t - 1$. We must show that, if $g, h \in G$ with $g \neq h$, then $\{x \mid g(x) = h(x)\}$ is the union of fewer than t blocks. By (iv), this set is a union of blocks. At the first stage at which both g_n and h_n are defined this set contains fewer than t blocks, by (iii) (or by assumption if $n = 0$); and the prescription of (b) guarantees that no further agreement occurs. Finally, the transitivity of G on bases of M (t-tuples from distinct blocks) is clear from (ii).

(ii)　　Now consider the case $L = \{0, 1, q, ..., q^{t-1}\}$, where q is a prime power. This time, let X be the point set of an affine space of countable dimension over GF(q), with affine basis $\{x_0, x_1, x_2, ...\}$. (We may take X to be a GF(q)-vector space of countable dimension, with $x_0 = 0$ and $\{x_1, x_2, ...\}$ a vector space basis). Again we construct pairs (G_n, m_n), where A_n is the affine span of $\{x_0, ..., x_{m_n}\}$. The conditions are:

(i)　　As before.

(ii)　　Each member of G_n is an affine bijection between affine subspaces of X; its domain and range contain A_n.

(iii) If $(y_0, ..., y_t)$ and $(z_0, ..., z_t)$ are affine independent (t+1)-tuples of elements of A_n, then a unique element of G_n carries y_i to z_i for $i = 0, ..., t$.

The construction is as before; the analogue of (iv) is the requirement that the transformations are affine (so that domains and ranges are affine subspaces). In (a), for each pair of tuples as in (iii) for which no transformation yet carries the first to the second, adjoin the unique affine transformation from $\langle y_0, ..., y_t \rangle$ to $\langle z_0, ..., z_t \rangle$ carrying y_i to z_i for $i = 0, ..., t$. The extension process in

(b) requires comment. Suppose that $g: U \to V$, where $g \in G_n$. Let $\dim(\langle U, A_{n+1} \rangle) - \dim(U) = r$. Let $x_{j+1}, \ldots, x_{j+r}$ be the first r basis vectors not previously used, and extend g to an affine transformation from $\langle U, A_{n+1} \rangle$ to $\langle V, x_{j+1}, \ldots, x_{j+r} \rangle$. The extension of the range is done similarly. The proof that the construction works is as before, noting that $\{x \mid g(x) = h(x)\}$ is an affine subspace.

References

Barlotti, A., & Strambach, K. (1984). K-transitive permutation groups and k-planes. Math. Z., **185**, 465–485.

Cameron, P.J., & Deza, M. (1979). On permutation geometries. J. London Math. Soc., (2), **20**, 373–386.

Cameron, P.J., Deza, M., & Frankl, P. (To appear). Sharp sets of permutations. J. Algebra.

Tits, J. (1952). Généralisation des groupes projectifs basée sur leurs propriétés de transitivité. Acad. Roy. Belgique Cl. Sci. Mem. **27**.

Tsuzuku, T. (1968). Transitive extensions of certain permutation groups of rank 3. Nagoya J. Math. **31**, 31–36.

Yoshizawa, M. (1979). On infinite four–transitive permutation groups. J. London Math. Soc. (2), **19**, 437–438.

Intersection and Containment Problems Without Size Restrictions

P. Frankl
C.N.R.S., Paris, France

0. Introduction

There has been a lot of progress recently in extremal finite set theory. Therefore we chose to treat only a small area of this topic. There is a recent paper by Füredi [Fü3] treating covering and matching type extremal problems. Intersection theorems were recently reviewed by the author, see [F2]. For an older but more general survey see [K].

Throughout this paper X will denote an n-element finite set. Usually, $X = \{1,2,\ldots,n\}$. A family, $\mathscr{F}$ is simply a subset of the power set 2^X.

Let us recall two of the best known results in extremal set theory.

Theorem 1. (Sperner [S]). Suppose that $\mathscr{F} \subset 2^X$ satisfies $F \not\subset F'$ for all distinct $F, F' \in \mathscr{F}$. Then

$$|\mathscr{F}| \leqslant \left(\left\lfloor \frac{n}{2} \right\rfloor \right) \text{ holds .}$$

Theorem 2. (Erdös, Ko and Rado [EKR]). Suppose that $n \geqslant 2k$, $k \geqslant 2$ and $\mathscr{F}$ is a family of k-element subsets of X satisfying $F \cap F' \neq \varnothing$ for all $F, F' \in \mathscr{F}$. Then

$$|\mathscr{F}| \leqslant \binom{n-1}{k-1} \text{ holds .}$$

Actually, Hilton and Milner [HM] showed that for $n > 2k$ the only way to achieve equality in Theorem 2 is to take all k-element subsets of X containing some fixed element.

Let us introduce the notation

$$\binom{X}{k} = \{\mathscr{A} \subset X : |\mathscr{A}| = k\} .$$

Also, for a family $\mathscr{F} \subset 2^X$ define

$$\sigma(\mathcal{F}) = \{G \subset X : \exists\, F \in \mathcal{F},\, G \subset F,\, |F - G| = 1\} \text{ and}$$

$$\sigma^{-1}(\mathcal{F}) = \{G \subset X : \exists\, F \in \mathcal{F},\, G \supset F,\, |G - F| = 1\}\,.$$

Sperner's theorem can be derived using the following simple result.

Proposition 3. ([S]) Suppose that $\varnothing \neq \mathcal{F} \subset \binom{X}{k}$. Then

$$|\sigma(\mathcal{F})| \big/ |\mathcal{F}| \geqslant \binom{n}{k-1} \big/ \binom{n}{k} \text{ and}$$

$$|\sigma^{-1}(\mathcal{F})| \big/ |\mathcal{F}| \geqslant \binom{n}{k+1} \big/ \binom{n}{k} \text{ hold}\,.$$

Moreover, equality holds if and only if $\mathcal{F} = \binom{X}{k}$.

Two further definitions. For $i \in X$ set

$$\mathcal{F}(i) = \{F - \{i\} : i \in F \in \mathcal{F}\} \text{ and}$$

$$\mathcal{F}(\bar{\imath}) = \{F : i \notin F \in \mathcal{F}\}\,.$$

For $0 \leqslant k \leqslant n$ set $f_k = |\{F \in \mathcal{F} : |F| = k\}|$, i.e., $f_k = |\mathcal{F} \cap \binom{X}{k}|$.

We shall also write $\mathcal{A} = \mathcal{A}_1 \oplus \cdots \oplus \mathcal{A}_t$ to denote that $\mathcal{A}_1, \ldots, \mathcal{A}_t$ partition $\mathcal{A}$, i.e., $\mathcal{A}_i \cap \mathcal{A}_j = \varnothing$ and $\mathcal{A} = \mathcal{A}_1 \cup \cdots \cup \mathcal{A}_t$.

Suppose that $\mathcal{F} \subset \binom{X}{k}$ and that for some $X = X_1 \oplus \cdots \oplus X_k$ $\mathcal{F}$ satisfies $|F \cap X_i| = 1$ for all $F \in \mathcal{F}$ and $1 \leqslant i \leqslant k$. Then $\mathcal{F}$ is called k-partite. The following important result can be proved by an averaging argument.

Theorem 4. (Erdös, Kleitman [EK2]). Suppose that $\mathcal{F} \subset \binom{X}{k}$. Then there exists $\mathcal{F}_0 \subset \mathcal{F}$, such that $\mathcal{F}_0$ is k-partite and satisfies $|\mathcal{F}_0| \big/ |\mathcal{F}| \geqslant k!/k^k$.

1. The Kleitman Lemma

Let us first recall that a downset (upset) on X is a family $\mathcal{F} \subset 2^X$ so that $G \subset F$ ($F \subset G \subset X$), $F \in \mathcal{F}$ imply $G \in \mathcal{G}$. Also $p(\mathcal{F}) = |F|/2^{|X|}$.

We start with a simple but very important result due to Kleitman [Kl1].

Lemma 1.1. Suppose $\mathcal{F}$ and $\mathcal{G}$ are both upsets on X. Then

$$p(\mathcal{F} \cap \mathcal{G}) \geqslant p(\mathcal{F}) p(\mathcal{G}) \tag{1.1}$$

holds.

Remark. This result says that considering a uniformly distributed random variable, ξ on 2^X, the events $\xi \in \mathcal{F}$ and $\xi \in \mathcal{G}$ are positively correlated.

Proof. (1.1) holds trivially if $|X| = 1$. Apply induction. Consider the families $\mathcal{F}(1)$, $\mathcal{F}(\bar{1})$, $\mathcal{G}(1)$, $\mathcal{G}(\bar{1})$. Clearly, they are all upsets; $|\mathcal{F}(1)| \geqslant |\mathcal{F}(\bar{1})|$, $|\mathcal{G}(1)| \geqslant |\mathcal{G}(\bar{1})|$ and $|\mathcal{F} \cap \mathcal{G}| = |\mathcal{F}(1) \cap \mathcal{G}(1)| + |\mathcal{F}(\bar{1}) \cap \mathcal{G}(\bar{1})|$. Using the above inequalities and the induction hypothesis we infer

$$p(\mathcal{F} \cap \mathcal{G}) = |\mathcal{F} \cap \mathcal{G}|/2^{|X|} = \frac{1}{2}\left[\frac{|\mathcal{F}(1) \cap \mathcal{G}(1)|}{2^{|X|-1}} + \frac{|\mathcal{F}(\bar{1}) \cap \mathcal{G}(\bar{1})|}{2^{|X|-1}}\right] = \frac{1}{2}(p(\mathcal{F}(1) \cap \mathcal{G}(1))$$

$$+ p(\mathcal{F}(\bar{1}) \cap \mathcal{G}(\bar{1}))) \geqslant \frac{1}{2}(p(\mathcal{F}(1))p(\mathcal{G}(1)) + p(\mathcal{F}(\bar{1}))p(\mathcal{G}(\bar{1}))) \geqslant \frac{1}{4}(p(\mathcal{F}(1)) + p(\mathcal{F}(\bar{1})))$$

$$\cdot (p(\mathcal{G}(1)) + p(\mathcal{G}(\bar{1}))) = p(\mathcal{F}) p(\mathcal{G}). \quad \blacksquare$$

Corollary 1.2. If $\mathcal{F}$ is an upset and $\mathcal{H}$ is a downset on X, then we have

$$p(\mathcal{F} \cap \mathcal{H}) \leqslant p(\mathcal{F}) p(\mathcal{H}). \tag{1.2}$$

Proof. Define $\mathcal{G} = 2^X - \mathcal{H}$. Then $\mathcal{G}$ is an upset, $p(\mathcal{G}) = 1 - p(\mathcal{H})$. Also $p(\mathcal{F}) = p(\mathcal{F} \cap \mathcal{H}) + p(\mathcal{F} \cap \mathcal{G})$. Thus using (1.1) we infer

$$p(\mathcal{F} \cap \mathcal{H}) = p(\mathcal{F}) - p(\mathcal{F} \cap \mathcal{G}) \leqslant p(\mathcal{F}) - p(\mathcal{F})p(\mathcal{G}) = p(\mathcal{F}) p(\mathcal{H}). \quad \blacksquare$$

The first and easiest theorem in extremal set theory says that if $\mathcal{F} \subset 2^X$ is intersecting (i.e., $F \cap F' \neq \varnothing$ holds for all $F, F' \in \mathcal{F}$) then $|\mathcal{F}| \leqslant 2^{|X|-1}$. Going over to complements this says that if $\mathcal{G} \subset 2^X$, and $G \cup G' \neq X$ holds for all $G, G' \in \mathcal{G}$ then $|\mathcal{G}| \leqslant 2^{|X|-1}$. What happens if we put these two restrictions simultaneously?

Theorem 1.3. If $\mathcal{A} \subset 2^X$ satisfies $A \cap A' \neq \varnothing$ and $A \cup A' \neq X$ for all $A, A' \in \mathcal{A}$ then

$$|\mathscr{A}| \leqslant 2^{n-2}. \tag{1.3}$$

Remark. This theorem, which is an easy consequence of the Kleitman lemma was proved independently by several people, e.g., Daykin and Lovász [DL], Marica and Schönheim [MS].

Proof. Consider $\mathscr{A}^* = \{B \subset X : \exists A \in \mathscr{A}, A \subset B\}$ and $\mathscr{A}_* = \{B : \exists A \in \mathscr{A}, B \subset A\}$. Clearly, $\mathscr{A}^*$ is an upset and A_* is a downset. By definition $A \cap A' \neq \varnothing$ for $A, A' \in \mathscr{A}^*$ and $A \cup A' \neq X$ for $A, A' \in \mathscr{A}_*$. Thus $p(\mathscr{A}^*) \leqslant \frac{1}{2}$, $p(\mathscr{A}_*) \leqslant \frac{1}{2}$. Now applying (1.2) we infer $p(\mathscr{A}) \leqslant p(\mathscr{A}^* \cap \mathscr{A}_*) \leqslant \frac{1}{4}$, i.e., $|\mathscr{A}| \leqslant 2^{n-2}$. $\blacksquare$

Remark. There are lots of extremal families. Suppose $X = X_1 \cup X_2$, $X_1 \cap X_2 = \varnothing$, $\mathscr{A}_1 \subset 2^{X_1}$, $\mathscr{A}_2 \subset 2^{X_2}$, $p(\mathscr{A}_1) = p(\mathscr{A}_2) = \frac{1}{2}$, $\mathscr{A}_1$ is intersecting, and $\mathscr{A}_2$ satisfies $A \cup A' \neq X_2$ for all $A, A' \in \mathscr{A}_2$. Define $\mathscr{A} = \mathscr{A}_1 \times \mathscr{A}_2 = \{A_1 \cup A_2 : A_1 \in \mathscr{A}_1, A_2 \in \mathscr{A}_2\}$. Then $p(\mathscr{A}) = \frac{1}{4}$ and $\mathscr{A}$ satisfies the assumptions of the theorem. One can show that all extremal families can be obtained in this way.

Theorem 1.4. (Kleitman [Kl1]) Suppose $\mathscr{F}_1, ..., \mathscr{F}_t$ are intersecting families, $\mathscr{F}_i \subset 2^X$. Then

$$p(\mathscr{F}_1 \cup \cdots \cup \mathscr{F}_t) \leqslant 1 - \frac{1}{2^t}.$$

Proof. Apply induction on t. For $t = 1$ the statement reduces to $|\mathscr{F}_1| \leqslant 2^{|X|-1}$. Set $\mathscr{G} = \mathscr{F}_2 \cup \cdots \cup \mathscr{F}_t$. Then by the induction assumption $p(\mathscr{G}) \leqslant 1 - \frac{1}{2^{t-1}}$.

By (1.1) $p(\mathscr{F}_1 \cap \mathscr{G}) \geqslant p(\mathscr{F}_1) p(\mathscr{G})$. Hence

$$p(\mathscr{F}_1 \cup \cdots \cup \mathscr{F}_t) = p(\mathscr{F}_1 \cup \mathscr{G}) = p(\mathscr{F}_1) + p(\mathscr{G}) - p(\mathscr{F}_1 \cap \mathscr{G}) \leqslant p(\mathscr{F}_1) + p(\mathscr{G}) - p(\mathscr{F}_1) p(\mathscr{G})$$

$$= 1 - (1 - p(\mathscr{F}_1))(1 - p(\mathscr{G})) \leqslant 1 - \left[1 - \frac{1}{2}\right]\left[1 - \left[1 - \frac{1}{2^{t-1}}\right]\right] = 1 - \frac{1}{2^t}. \quad \blacksquare$$

Remark. To achieve equality in the theorem take t pairwise disjoint subsets $X_1, ..., X_t$ of X and let $\mathscr{A}_i \subset 2^{X_i}$ be intersecting of maximal size, i.e., $|\mathscr{A}_i| = 2^{|X_i|-1}$. Define $\mathscr{F}_i = \{F \subset X : F \cap X_i \in \mathscr{A}_i\}$. Then $|\mathscr{F}_1 \cup \cdots \cup \mathscr{F}_t| = 2^n\left[1 - \frac{1}{2^t}\right]$, as desired. One can show that there are no other extremal constructions.

2. No k Pairwise Disjoint Sets

If $A_1, ..., A_k$ are pairwise disjoint subsets of the n-set X, then one of them has size at most $\left\lfloor \dfrac{n}{k} \right\rfloor$. Thus $\left(\genfrac{}{}{0pt}{}{X}{> \left\lfloor \frac{n}{k} \right\rfloor} \right) = \left\{ A \subset X : |A| > \left\lfloor \dfrac{n}{k} \right\rfloor \right\}$ provides a large family without k pairwise disjoint sets. If $k=2$, n odd then it has size 2^{n-1}, therefore it is optimal. For $k \geq 3$ it contains all sets whose size is around $\dfrac{n}{2}$, thus $2^n(1-o(1))$ sets. If $n \not\equiv -1 \pmod{k}$ then one can add some of the $\left\lfloor \dfrac{n}{k} \right\rfloor$-element subsets to $\left(\genfrac{}{}{0pt}{}{X}{> \left\lfloor \frac{n}{k} \right\rfloor} \right)$ without introducing k pairwise disjoint sets. E.g., for n divisible by k, that is $n=mk$ one can add $\left(\genfrac{}{}{0pt}{}{X-\{1\}}{m} \right)$. In general, if $n=mk+i$, $1 \leq i \leq k-2$ any family of m-subsets without $(k-i)$ pairwise disjoint ones. In fact, suppose $A_1, ..., A_k$ are pairwise disjoint in the extended family and say $A_1, ..., A_j$ are the m-sets. Then $j < k-i$, therefore there are at least $i+1$ sets of size at least $m+1$. This yields $|A_1| + \cdots + |A_k| \geq (k-i-1)m + (i+1)(m+1) = km+i+1 > n$, a contradiction.

The main result of this section, which is due to Kleitman [Kl2], is that these constructions are best possible for $n=mk-1$ and $n=mk$.

Theorem 2.1. Suppose $\mathscr{F} \subset 2^X$, $k \geq 3$ and there are no k pairwise disjoint sets in $\mathscr{F}$. Then for

(a) $|X| = mk$

$$|\mathscr{F}| \leq \sum_{j=m+1}^{mk} \binom{mk}{j} + \frac{k-1}{k} \binom{mk}{k}, \tag{2.1}$$

moreover, in case of equality for some $y \in X$ we have $\mathscr{F} = \{F \subset X : |F \cap (X-\{y\})| \geq m\}$.

(b) $|X| = mk-1$

$$|\mathscr{F}| \leq \sum_{j=m}^{mk-1} \binom{mk-1}{j}, \tag{2.2}$$

moreover in case of equality $\mathscr{F} = \left(\genfrac{}{}{0pt}{}{X}{\geq m} \right)$.

Proof of (a). Suppose $n \geqslant m_1 + m_2 + \cdots + m_k$. Consider all ordered partial partitions π: $A_1 \oplus \cdots \oplus A_k \subset X$ with $|A_i| = m_i$. Clearly each m_i-subset is the i-th set in such a partition in a proportion of $1 / \binom{n}{m_i}$ of all such partitions. Also, from each partition there is at least one set which lies outside $\mathscr{F}$. Denoting by g_t the number of t-element sets outside $\mathscr{F}$, i.e., $g_t = \binom{n}{t} - f_t$, then this leads to

$$\sum_{1 \leqslant i \leqslant k} g_{m_i} / \binom{n}{m_i} \geqslant 1. \tag{2.3}$$

We can and need to make this statement a little more accurate. Denote by $x_j(\pi)$ the proportion of partial partitions of type π, i.e., i'th set of size m_i, from which $\mathscr{F}$ misses exactly j sets.

Lemma 2.2.

(i) $\quad x_0(\pi) = 0$

(ii) $\quad \sum_{j=1}^{k} x_j(\pi) = 1,$

(iii) $\quad \sum_{j=1}^{k} j\, x_j(\pi) = \sum_{1 \leqslant i \leqslant k} g_{m_i} / \binom{n}{m_i}.$

Proof of the Lemma. (i) and (ii) are direct consequences of the definitions. To obtain (iii) note that both sides count the average over all partitions of the number of sets outside $\mathscr{F}$: this is trivial for the LHS. On the RHS it is done via adding up the proportions in which the i'th set is missing. ∎

From now on suppose $n = mk$ and also that $\mathscr{F}$ is an upset. Let us apply the lemma to the partitions of type π_e: $n = m + m + \cdots + m$ (equipartitions). We obtain

$$\sum_{j=1}^{k} j\, x_j(\pi_e) = k g_m / \binom{n}{m}, \quad \text{or using (ii)}$$

$$g_m = \frac{1}{k} \binom{mk}{m} + \frac{1}{k} \binom{mk}{m} \sum_{j=2}^{k} (j-1)\, x_j(\pi_e). \tag{2.4}$$

Applying (2.3) to partitions of type $(m+1) + \cdots + (m+1) + (m+1-k-j)$ (i.e., partial partitions with $(k-1)$ sets of size $m+1$ and one of $(m+1-k-j))$, we obtain

$$g_{m+1-k-j} + (k-1)g_{m+1}\binom{n}{m+1-k-j}\bigg/\binom{n}{m+1} \geqslant \binom{n}{m+1-k-j}. \qquad (2.5)$$

We need one more inequality $(j < k)$:

$$g_{m-j} + \frac{(k-1)g_{m+1}\binom{n}{m-j}}{\binom{n}{m+1}} \geqslant \binom{n}{m-j} - \sum_{r=j+1}^{k}\left[1 - \frac{r}{k}\right]x_r\,(\pi_e)\binom{n}{m-j}. \qquad (2.6)$$

Proof of (2.6). Let us consider partial partitions into k sets of type π_j: $(m-j)+m+m+\cdots+m$. If $A_1 \oplus A_2 \oplus \cdots \oplus A_k$ is such a partition and $B_1 = X - (A_2 \oplus \cdots \oplus A_k)$, then $B_1 \oplus \cdots \oplus A_k$ is an equipartition associated with the partial partition. Among the equipartitions a proportion $x_r\,(\pi_e)$ has r missing sets from $\mathscr{F}$, and thus by symmetry out of those $\frac{r}{k}x_r\,(\pi_e)$ have their first set B_1 not in $\mathscr{F}$. Then $A_1 \notin \mathscr{F}$, too. Summing these over $r > j$ we obtain already that in at least a proportion of $\sum_{j<r\leqslant k}\frac{r}{k}x_r\,(\pi_e)$ of the partial partitions of type π_j $A_1 \notin \mathscr{F}$ holds. We have to add still the proportion corresponding to those partial partitions where the associated equipartition has at most j missing sets.

Let $A_1 \oplus \cdots \oplus A_k$ be such a partition and suppose $A_1 \in \mathscr{F}$, and say $A_2, ..., A_{r+1} \notin \mathscr{F}$.

Consider a bipartite graph with vertex set $\mathscr{A} = \{A_2, ..., A_{r+1}\}$ in one part and the elements of $B_1 - A_1$ in the other and an edge if and only if $(A_j \cup \{b\}) \in \mathscr{F}$.

Clearly this graph cannot contain a matching from $\mathscr{A}$ to $B_1 - A_1$ since that would give sets $B_2,...,B_{r+1}$ in $\mathscr{F}$ which along with $A_1, A_{r+2},...,A_k$ would be k pairwise disjoint sets. Using König's theorem (*cf.* [L]) we can find $r-1$ vertices say $A_{i_1},...,A_{i_t}$ and $b_{t+1},...,b_{r-1}$ covering all the edges. Consequently there at least $(r-t)(j+t+1-r) \geqslant j$ edges in the complement. That is, we obtain at least j $(m+1)$-sets lying outside of $\mathscr{F}$. This represents a proportion $\frac{1}{k-1}$ of the $(m+1)$-sets obtainable in this way. Therefore in at least a proportion of $\sum_{r=1}^{j} x_r\,(\pi_e) - (k-1)g_{m+1}\binom{n}{m+1}$ of partial partitions, whose associated equipartition misses at most j sets, the $(m-j)$-set is missing from $\mathscr{F}$. This gives altogether

$$\frac{g_{m-j}}{\binom{n}{m-j}} \geqslant \sum_{j<r\leqslant k} \frac{r}{k} x_r(\pi_e) + \sum_{r=1}^{j} x_r(\pi_e) - \frac{(k-1)g_{m+1}}{\binom{n}{m+1}},$$

which is equivalent to (2.6).

Now we want to sum up the inequalities (2.5) for $0 \leqslant j \leqslant m+1-k$, (2.6) for $j=1,...,k-1$ and (2.4). Then the total coefficient of g_{m+1} will be:

$$(k-1) \sum_{j=0}^{m-1} \binom{n}{j} \Big/ \binom{n}{m+1} \stackrel{\text{def}}{=} c_{m+1}.$$

Claim $c_{m+1} < 1$.

It is easy to check that for $j < m-1$ one has

$$\binom{n}{j} \Big/ \binom{n}{j+1} \leqslant \binom{n}{m-2} \Big/ \binom{n}{m-1} = \frac{m-1}{m(k-1)+2}.$$

Consequently,

$$\sum_{j=0}^{m-1} \binom{n}{j} < \frac{m(k-1)+2}{m(k-2)+3} \binom{n}{m-1} < \frac{m(k-1)+1}{m+1} \binom{n}{m-1} = \binom{n}{m+1} \Big/ (k-1), \text{ proving the claim.}$$

Let us count the overall coefficient of $x_r(\pi_e)$, $2 \leqslant r \leqslant k$. It is:

$$\frac{1}{k} \binom{n}{m} (r-1) - \sum_{j=1}^{r-1} \frac{k-r}{k} \binom{n}{m-j}.$$

One can show as above $\sum_{j=1}^{r-1} \binom{n}{m-j} < \frac{1}{k-2} \binom{n}{m}$. Therefore $r-1 \geqslant \frac{k-r}{k-2}$ implies that this

coefficient is positive. Thus we obtain from the sum of the inequalities

$$\sum_{j=0}^{m} g_j + c_{m+1} g_{m+1} \geqslant \sum_{j=0}^{m-1} \binom{n}{j} + \frac{1}{k} \binom{n}{m}.$$

Consequently,

$$|\mathscr{F}| \leqslant 2^n - \sum_{j=0}^{m+1} g_j \leqslant \frac{k-1}{k} \binom{mk}{m} + \sum_{j=m+1}^{mk} \binom{mk}{j}, \tag{2.7}$$

as desired. ∎

In order to have equality in (2.7) we must have $g_j = 0$ for $j > m+1$ and $g_{m+1} = 0$ follows from $c_{m+1} < 1$. Also, the positivity of the overall coefficients of $x_j(\pi_e)$ implies $x_j(\pi_e) = 0$ for $j \geq 2$. Thus $f_m = \binom{n}{m} - g_m = \binom{n}{m} - \frac{1}{k}\binom{n}{m} = \binom{n-1}{m}$ follows from (2.4). Now $x_1(\pi_e) = 1$ implies that $\binom{s}{m} - \mathscr{F}_m$ is an intersecting family of size $\binom{n-1}{m-1}$. Since $k \geq 3$ all its members contain some fixed element, say y. Then $\mathscr{F}_m = \{F \in \binom{X}{m} : y \notin F\}$.

Using (2.7) again we infer $f_j = 0$ for $j < m$. (a) is proved.

Proof of (b). First note that

$$\sum_{j=m+1}^{mk} \binom{mk}{j} + \frac{k-1}{k}\binom{mk}{m} = 2\sum_{j=m}^{mk-1}\binom{mk-1}{j},$$

(to see this, just write $\binom{mk}{j} = \binom{mk-1}{j} + \binom{mk-1}{j-1}$ for $j > m$ and $\frac{k-1}{k}\binom{mk}{m} = \binom{mk-1}{m}$).

Suppose now $\mathscr{F} \subset 2^{\{1,2,\ldots,mk-1\}}$, $\mathscr{F}$ contains no k pairwise disjoint sets. Define $\mathscr{\bar F} = \{E \subset \{1,2,\ldots,mk\} : (E - \{mk\}) \in \mathscr{F}\}$. Clearly, $|\mathscr{\bar F}| = 2|\mathscr{F}|$ and $\mathscr{\bar F}$ contains no k pairwise disjoint members either. Thus (2.2) follows from (2.1).

Suppose now that $\mathscr{F}$ attains equality in (2.2). Then $\mathscr{\bar F}$ attains equality in (2.1), therefore $f'_j = 0$ for $j < m$. Thus $f_j = 0$ for $j < m$. Consequently $\mathscr{F} = \binom{\{1,\ldots,mk-1\}}{\geq m}$. $\blacksquare$

For a quantitative version of Theorem 2.1 see [AF].

3. No k Disjoint Sets Along with Their Union

Let us write $|X| = n$ in the form $n = (k+1)t + i$, $0 \leq i \leq k$. Taking $\mathscr{A}(t+1) = \{F \subset X : t+1 \leq |F| < k(t+1)\}$, we obtain a fairly large family without k pairwise disjoint members whose union is in the family. For $k \geq 3$ and $i \leq k-3$ one can do little better, namely fixing $Y \in \binom{X}{k-i-2}$ and setting

$$\mathscr{F}^*_{(t+1)} = \left\{F \in \binom{X}{t} : F \cap Y \neq 0\right\} \cup \{F \subset X : t+1 \leq |X| \leq kt+i+1\}.$$

The aim of this section is to prove the next theorem which was proved by Kleitman [K13] for $k=2$ and by Frankl [F1] for general k.

Theorem 3.1. Suppose $\mathscr{F} \subset 2^X$, $|X| = n = k(t+1) + k - 1$ and $\mathscr{F}$ contains no k disjoint members along with their union. Then $|\mathscr{F}| \leq |\mathscr{F}(t+1)|$ and $|\mathscr{F}| = |\mathscr{F}(t+1)|$ implies $\mathscr{F} = \mathscr{F}(t+1)$.

For the proof we need a lemma which is valid for any n (i.e., not only for $n \equiv -2 \pmod{k+1}$). Denote by g_j the number of j-subsets not contained in $\mathscr{F}$.

Lemma 3.2. Suppose $n = t_1 + t_2 + \cdots + t_{k+1}$ then

$$\sum_{1 \leq i \leq k+1} \frac{g_{t_i} + g_{n-t_i}}{\binom{n}{t_i}} \geq 2. \tag{3.1}$$

Proof. Consider ordered partitions of X of the form $X = A_1 \oplus A_2 \oplus \cdots \oplus A_{k+1}$ with $|A_i| = t_i$. Note that the LHS of (3.1) counts the average number of sets lying outside $\mathscr{F}$ among $A_1, \ldots, A_{k+1}$, $X - A_1, \ldots, X - A_{k+1}$. Since this number is easily seen to be always at least 2, the lemma is proved. ∎

Let us note that in case of equality always exactly 2 sets have to lie outside $\mathscr{F}$.

Proof of the Theorem. Suppose $n = (k+1)t + k - 1$.

Applying the lemma to partitions with $|A_1| = |A_2| = t$ and $|A_3| = \cdots = |A_{k+1}| = t+1$ we infer

$$g_t + g_{n-t} + \frac{(k-1)\binom{n}{t}}{2\binom{n}{t+1}}(g_{t+1} + g_{n-t-1}) \geq \binom{n}{t}. \tag{3.2}$$

Using (3.1) with $t_1 = \cdots = t_j = t+2+i$, $t_{j+1} = \cdots = t_k = t+1+i$, $t_{k+1} = t-1-ki-j$, we obtain

$$g_{t-1-ki-j} + g_{n-(t-1-ki-j)} + \frac{(k-j)\binom{n}{t-1-ki-j}}{\binom{n}{t+1+i}} (g_{t+i+1} + g_{n-t-i-1})$$

$$+ \frac{j\binom{n}{t-1-ki-j}}{\binom{n}{t+2+i}} (g_{t+i+2} + g_{n-t-i-2}) \geqslant 2\binom{n}{t-1-ki-j}, \qquad (3.3)$$

$$\text{for} \quad 0 \leqslant j \leqslant = k-1, \quad 0 \leqslant i, \quad t-1-ki-j \geqslant 0.$$

We want to sum up these inequalities along with (3.2). Let c_r be the total coefficient of g_r.

Claim: $c_r < 1$ for $t+1 \leqslant r \leqslant = n-t-1$, $c_r = 1$ otherwise.

Only the first part needs a proof.

One has $c_{t+1} = c_{n-t-1} = \dfrac{(k-1)\binom{n}{t}}{2\binom{n}{t+1}} + \displaystyle\sum_{j=0}^{k-1} \dfrac{\binom{n}{t-1-j}}{\binom{n}{t+1}}.$

Since $\quad \binom{n}{t} / \binom{n}{t+1} = \dfrac{t+1}{kt+k-1},$ $\qquad \dfrac{(k-1)\binom{n}{t}}{2\binom{n}{t+1}} < 1/2.$ $\qquad$ Also

$$\binom{n}{t-1} / \binom{n}{t+1} = \frac{(t+1)t}{k(t+1)(kt+k-1)} < \frac{1}{k^2}, \quad \text{yielding} \quad \sum_{j=0}^{k-1} \frac{\binom{n}{t-1-j}}{\binom{n}{t+1}} \leqslant \frac{k\binom{n}{t-1}}{\binom{n}{t+1}} \leqslant \frac{1}{k} \leqslant \frac{1}{2}.$$

Hence $c_{t+1} < 1$. The remaining inequalities can be proved in a similar but easier way.

Thus summing up (3.3) for $0 \leqslant j \leqslant k-1$ and $0 \leqslant i$, $t-1-ki-j \geqslant 0$ along with (3.2) yields

$$\sum_{0 \leqslant r \leqslant n} g_r \geqslant \sum_{0 \leqslant r \leqslant n} c_r g_r \geqslant \sum_{i=0}^{t-1} 2\binom{n}{i} + \binom{n}{t}.$$

We infer

$$|\mathscr{F}| = 2^n - \sum_{0 \leqslant r \leqslant n} g_r \leqslant \sum_{t+1=l}^{kt+k-1} \binom{n}{l} \quad \text{as desired.}$$

If we have equality we must have $g_r = 0$ for each r with $c_r < 1$, that is

$$\mathscr{F} \supset \{F \subset X : t+1 \leqslant |F| \leqslant kt+k-2\}. \qquad (3.4)$$

This trivially implies $t \leqslant |F| \leqslant kt+k-1$ for all $F \in \mathscr{F}$ and $f_t + f_{n-t} = \binom{n}{t}$. To finish the proof of the theorem we have to show $f_t = 0$. By (3.4) $\mathcal{F}_t$ cannot contain 2 disjoint t-element sets. Therefore $f_t < \binom{n}{t}$.

Consider $\sigma^{n-t}(\mathcal{F}_t) = \{G \in \binom{X}{n-t} : \exists F \in \mathcal{F}_t, F \subset G\}$. Clearly $\sigma^{n-t}(\mathcal{F}_t) \cap \mathcal{F}_{n-t} = \varnothing$ (any such G would be the disjoint union of F along with $(k-1)$ $(t+1)$-sets). If $\mathcal{F}_t \neq \varnothing$, then $|\sigma^{n-t}(\mathcal{F}_t)| > |\mathcal{F}_t|$ follows from $\mathcal{F}_t \neq \binom{X}{t}$ (by Proposition 3). Thus $|\mathcal{F}_t| + |\mathcal{F}_{n-t}| < \binom{n}{t}$ — and we have too few sets in $\mathcal{F}$. This final contradiction shows $\mathcal{F}_t = \varnothing$ and $\mathcal{F} = \mathcal{A}(t+1)$. $\blacksquare$

4. No Two Sets with Their Union

The problem investigated in this section is: what is $\max|\mathcal{F}|$ if there are no distinct $A, B, C \in \mathcal{F}$ with $A \cup B = C$. Clearly $\binom{X}{\lceil \frac{n}{2} \rceil}$ satisfies this condition. But it is not maximal. One can add to it

any $\mathcal{G} \subset \binom{X}{\lceil \frac{n}{2} \rceil - 1}$, satisfying $|G \triangle G'| > 2$ or equivalently $|G \cup G'| > \lceil \frac{n}{2} \rceil$ for all distinct

$G, G' \in \mathcal{G}$.

Suppose $X = \{1, ..., n\}$ and $1 \leqslant i \leqslant n$ is fixed. Define

$$\mathcal{G}(i) = \left\{ G \in \binom{X}{\lceil \frac{n-2}{2} \rceil} : \sum_{j \in G} j \equiv i \pmod{n} \right\}.$$

It is easy to see that $|G \triangle G'| > 2$ holds for $G, G' \in \mathcal{G}(i)$. Choose i so as to maximize $|\mathcal{G}(i)|$. Then $|\mathcal{G}(i)| \geqslant \frac{1}{n} \binom{n}{\lceil \frac{n-2}{2} \rceil}$. Define $\mathcal{G} = \mathcal{G}(i)$ and $\mathcal{H} = \binom{X}{\lceil \frac{n}{2} \rceil} \cup \mathcal{G}$. Then $\mathcal{H}$ satisfies the

conditions and $|\mathcal{H}| \geqslant \binom{n}{\lceil \frac{n}{2} \rceil} \left(1 + \frac{1 + o(1)}{n} \right)$. Our aim is to prove

Theorem 4.1. (Kleitman [Kl4]). Suppose $\mathcal{F} \subset 2^X$ and there are no distinct $A, B, C \in \mathcal{F}$ with $A \cup B = C$, then

$$|\mathscr{F}| \le \left[\binom{n}{\lfloor\frac{n}{2}\rfloor}\right] + \frac{2^n}{n+1} + O\left(\left[\binom{n}{\lfloor\frac{n}{4}\rfloor}\right]\right).$$

Proof. Suppose $\mathscr{F}$ satisfies the assumptions. First we prove some inequalities.

Proposition 4.2.

$$\sum_{j=k}^{2k} f_j / j\binom{n}{j} \le \frac{1}{k}, \quad \text{for } 1 \le k \le \frac{n}{2}.$$

Proof of the proposition. Consider chains of type $A_k \subset A_{k+1} \subset \cdots \subset A_{2k}$ with $|A_j| = j$, $k \le j \le 2k$. Clearly, a given j-set B satisfies $B = A_j$ in a proportion $1/\binom{n}{j}$ of all such chains.

In each chain put a star to the smallest member which belongs to $\mathscr{F}$, if there is any. Clearly, at most one member is starred. Fix $F \in \mathscr{F}$ with $|F| = j$. Define

$$\mathscr{A}(F) = \left\{ A \in \binom{X}{k} : \exists F(A) \in \mathscr{F}, \ A \subset F(A) \underset{\ne}{\subseteq} F \right\}.$$

Then clearly for $A, B \in \mathscr{A}(F)$ $A \cup B = F$ implies $F(A) \cup F(B) = F$, which is excluded. Thus $\mathscr{B}(F) \overset{\text{def}}{=} \{F - A : A \in \mathscr{A}(F)\}$ is an intersecting family of $(j-k)$-subsets of a j-set.

Since $j \le 2k$ implies $2(j-k) \le j$, we may apply the Erdös-Ko-Rado theorem (Theorem 2) and obtain $|\mathscr{A}(F)| \le \binom{j-1}{j-k-1}$. That is $\dfrac{|\mathscr{A}(F)|}{\binom{j}{k}} \le \dfrac{k}{j}$.

This entails that in at least a proportion $\dfrac{k}{j}$ of the chains in which it appears, F is starred. Since in each chain at most one member is starred,

$$\sum_{j=k}^{2k} \frac{k}{j} f_j / \binom{n}{j} \le 1$$

follows. ∎

To prove the theorem let us consider the inequalities of the proposition along with the trivial inequalities $f_j / \binom{n}{j} \le 1$ as a linear program with objective function $\sum_{j=0}^{n} f_j$.

By the duality theorem every solution of the dual program is an upper bound for $\sum_{j \geq 0} f_j$. The

dual program has variables y_k, z_j, $0 \leq k \leq \frac{n}{2}$, $0 \leq j \leq n$. It reads:

$$\sum_{k=j/2}^{\min\left\{j, \left\lfloor \frac{n}{2} \right\rfloor\right\}} \frac{y_k}{j \binom{n}{j}} + \frac{z_j}{\binom{n}{j}} \geq 1$$

minimize $\sum_{j=0}^{n} z_j + \sum \frac{y_k}{k}$.

One can easily check that $z_j = 0$, $y_{2k+1} = (2k+1) \binom{n}{2k+1} - 2k \binom{n}{2k} + y_k$,

$y_{2k} = 2k \binom{n}{2k} - (2k-1) \binom{n}{2k-1}$ defines a solution of this program. The corresponding value of the

objective function is

$$\sum_{k=1}^{\left\lfloor \frac{n}{2} \right\rfloor} \frac{y_k}{k} = \binom{n}{\left\lfloor \frac{n}{2} \right\rfloor} + \sum_{j=\left\lfloor \frac{n}{4} \right\rfloor}^{\left\lfloor \frac{n-2}{2} \right\rfloor} \frac{1}{j+1} \binom{n}{j} + O\left(\binom{n}{\left\lfloor \frac{n}{4} \right\rfloor} \right) \leq \binom{n}{\left\lfloor \frac{n}{2} \right\rfloor} + \frac{2^n}{n+1} + O\left(\binom{n}{\left\lfloor \frac{n}{4} \right\rfloor} \right).$$

Since this is an upper bound for $|\mathscr{A}| = \sum_{i=0}^{n} f_i$, the proof is complete. ∎

Katona and Tarján [KT] considered the following problem: What is $\max |\mathscr{A}|$ if $\mathscr{F}$ contains no 3

sets F, G, H with $F \supset G$, $F \supset H$ or $F \subset G$, $F \subset H$ that is every member of $\mathscr{F}$ is contained in at

most one other member and contains at most one other member of $\mathscr{F}$. They show that

$$\max |\mathscr{A}| = \begin{cases} \binom{n}{\frac{n}{2}} & \text{if } n \text{ is even} \\ 2 \binom{n-1}{\frac{n-1}{2}} & \text{if } n \text{ is odd} \end{cases}.$$

Their construction for $n = 2t + 1$ is:

$$\mathscr{G} = \{F \subset X : |F \cap (X - \{1\})| = t\}.$$

Then $|\mathscr{G}| = 2 \binom{n-1}{\frac{n-1}{2}} = \left(1 + \frac{1}{n}\right) \binom{n}{\left\lfloor \frac{n}{2} \right\rfloor}$, thus it provides an alternative construction for large families satisfying the assumptions of Theorem 4.1. In the case $n = 2t$ define

$$\mathcal{G} = \{G \subset X : 1 \in G, |G| = t+1 \text{ or } 1 \notin G, |G| = t-1\}.$$

Suppose now that for all distinct members $A, B, C \in \mathcal{F}$ neither $A \cap B = C$ nor $A \cup B = C$ holds. This is a weaker condition than the one mentioned above. Recently Kleitman [K15] proved that subject to this restriction $\max |\mathcal{F}|$ for $\mathcal{F} \subset 2^X$ is $2\left[\begin{array}{c} n-1 \\ \frac{n-1}{2} \end{array}\right]$ for n odd and $\left[\begin{array}{c} n \\ \frac{n}{2} \end{array}\right] + 2$ for n even $(n > 20)$. To get the extra members in the even case just take $\mathcal{F} = \mathcal{G} \cup \{1\} \cup (X - \{1\})$.

5. Families in which no Set is Covered by the Union of Two Others

For a family $\mathcal{F} \subset 2^X$ and a subset A of a member F of $\mathcal{F}$, we call A an *own subset* of F if $A \subset G \in \mathcal{F}$ implies $G = F$.

The next proposition is obvious.

Proposition 5.1. Suppose $\mathcal{F} \subset 2^X$ contains no three sets F_0, F_1, F_2 with $F_0 \subset F_1 \cup F_2$. Suppose $A \cup B = F \in \mathcal{F}$. Then at least one out of A and B is an own subset of F. ∎

Corollary 5.2. Suppose $\mathcal{F} \subset \binom{X}{k}$ and $\mathcal{F}$ contains no three sets F, G, H with $F \subset G \cup H$. Then

$$|\mathcal{F}| \leqslant 2 \left[\begin{array}{c} n \\ \left\lceil \frac{k}{2} \right\rceil \end{array}\right] \bigg/ \left[\begin{array}{c} k \\ \left\lceil \frac{k}{2} \right\rceil \end{array}\right]. \tag{5.1}$$

Proof. First consider the case $k = 2l$. By Proposition 5.1 each $F \in \mathcal{F}$ has at least $\frac{1}{2}\binom{2l}{l}$ own subsets of size l. Since the total number of own l-sets cannot exceed $\binom{n}{l}$, (5.1) follows. Suppose now $k = 2l+1$, $F \in \mathcal{F}$. It will be sufficient to show that F has at least $\frac{1}{2}\binom{2l+1}{l+1}$ own subsets of size $l+1$. Suppose for contradiction that some F has less than $\frac{1}{2}\binom{2l+1}{l+1}$ own $(l+1)$-sets. Then by Proposition 5.1 it has more than $\frac{1}{2}\binom{2l+1}{l+1} - \frac{1}{2}\binom{2l+1}{l}$ own subsets of size l. Let $\mathcal{A} = \{A_1, ..., A_m\}$ be their collection. By definition $\sigma^{-1}(\mathcal{A}) = \{B \in \binom{F}{l+1} : \exists A \in \mathcal{A}, A \subset B\}$ consists only of own $(l+1)$-subsets of F. In view of Proposition 3 $|\sigma^{-1}(\mathcal{A})| \geqslant |\mathcal{A}| > \frac{1}{2}\binom{2l+1}{l}$, the desired contradiction. ∎

Let us now look at the problem how one can construct families satisfying the condition in the title of the section. Suppose $\mathscr{S} \subset \binom{X}{k}$ and for all $S, S' \in \mathscr{S}$, $|S \cap S'| < t$ holds. Such a family is called a partial Steiner-system, $PS(n, k, t)$. Clearly any such $\mathscr{S}$ satisfies $|\mathscr{S}| \leqslant \binom{n}{t} / \binom{k}{t}$. If equality holds then $\mathscr{S}$ is called a Steiner-system $S(n, k, t)$. The existence problem of Steiner-systems is a very difficult question. Let us mention the following result of Rödl [R1]:

For every $k > t \geqslant 1$ there exists a $PS(n, k, t)$ $\mathscr{S}$ with

$$|\mathscr{S}| = (1 + o(1)) \binom{n}{t} / \binom{k}{t}.$$

Since any $PS(n, 2l-1, l)$ satisfies our union condition, there exist $(2l-1)$-uniform families of size $(1 + o(1)) \binom{n}{l} / \binom{2l-1}{l}$ without two sets whose unions covers a third.

Adjoining a fixed element to all the subsets we obtain a $2l$-uniform family of size $(1 + o(1)) \binom{n-1}{l} / \binom{2l-1}{l}$.

Theorem 5.3. (Erdős, Frankl, Füredi [EFF1]). Suppose $\mathscr{F} \subset \binom{X}{k}$ and there are no three distinct sets $F, G, H \in \mathscr{F}$ such that $F \subset G \cup H$. Then we have

$$\text{(a)} \quad k = 2l-1 \qquad\qquad |\mathscr{F}| \leqslant \binom{n}{l} / \binom{2l-1}{l}.$$

$$\text{(b)} \quad k = 2l \qquad\qquad |\mathscr{F}| \leqslant \binom{n-1}{l} / \binom{2l-1}{l}.$$

Proof. Suppose $\mathscr{F} \subset \binom{X}{2l-1}$ and $\mathscr{F}$ satisfies the assumptions. Then clearly so does $\mathscr{F}' = \{F \cup \{n+1\} : F \in \mathscr{F}\}$, $|\mathscr{F}'| = |F|$, $F' \subset \binom{X \cup \{n+1\}}{2l}$. Thus it is sufficient to prove (b).

However, to give the flavor of the proof, first we give a simpler argument proving (a) for $n \geqslant 3l-1$. Define a weight function $W_{(F,A)}: \binom{X}{l} \to \mathbb{R}^+$ for each pair $A \subset F \in \mathscr{F}$, $|A| = l$. Set

$$W_{(F,A)}(B) = \begin{cases} 1 & \text{for } B = A \\ 0 & \text{otherwise} \end{cases}$$

if A is an own subset of F and in the remaining case

$$W_{(F,A)}(B) = \begin{cases} \dfrac{1}{n-2l+1} & \text{if } |B|=l, \ B \cap F = F - A \\[2mm] 0 & \text{otherwise} \end{cases}$$

It is easy to see that in both cases $\displaystyle\sum_{B \in \binom{X}{l}} W_{(F,A)}(B) = 1$ holds. Thus we have

$$\sum_{F \in \mathscr{F}} \sum_{A \in \binom{F}{l}} \sum_{B \in \binom{X}{l}} W_{(F,A)}(B) = \binom{2l-1}{l} |\mathscr{F}|. \tag{5.2}$$

Claim 5.4. For all $B \in \binom{X}{l}$ we have

$$\sum_{\substack{(F,A) \\ A \subseteq F \in \mathscr{F}}} W_{(F,A)}(B) \leqslant 1. \tag{5.3}$$

Proof of the Claim. Let us note first that if $W_{(F,A)}(B) \neq 0$ for some $B \neq A$ then $F - A$ is an own subset of F, $F - A \subset B \not\subseteq F$, thus B is not contained in any $F' \in \mathscr{F}$ either. For such a B, $W_{(F,A)}(B) > 0$ is possible only if one of the l $(l-1)$-subsets of B, say C satisfies $A \cup C = F$, C is an own subset of F. That is C determines (F,A) uniquely. We obtain that for such B the LHS of (5.3) is upperbounded by $\dfrac{l}{n-2l+1} \leqslant 1$ for $n \geqslant 3l-1$. If B is contained in some $F \in \mathscr{F}$ then the LHS of (5.3) is 1 or 0 according whether B is an own subset of F or not, proving (5.3). ∎

Since there are only $\binom{n}{l}$ choices for B, (5.3) and (5.2) imply $\binom{2l-1}{l}|\mathscr{F}| \leqslant \binom{n}{l}$, which proves (a).

Let us now give the more complicated argument for (b). Let us call $B \in \binom{X}{l}$ *free* if B is not contained in any $F \in \mathscr{F}$.

If $T \subset F \in \mathscr{F}$, $|T|=l$ and for some $x \in (F-T)$ the set $T \cup \{x\}$ is not an own subset of F then by Proposition 5.1 $(F-T)-\{x\}$ is an own subset of F. Consequently, for all $y \in (X-F)$, $(F-T-\{x\}) \cup \{y\}$ is free. We call it a free subset associated with the pair (F,T). Let us denote by $\mathscr{A}(F,T)$ the collection of all such free sets. Formally

$$\mathcal{A}(F,T) = \{A \in \binom{X}{l} : A \cap T = \varnothing, \ |A \cap F| = l-1, \ |\mathcal{F}(F-A)| > 1\}.$$

Claim 5.5.

$$|\mathcal{A}(F,T)| = |X-F||\{x \in F-T : |\mathcal{F}(T \cup \{x\})| > 1\}|. \quad \blacksquare$$

If $|\{F \in \mathcal{F} : T \subset F\}| > (n-1)/l$ then we have

$$\sum_{T \subset F \in \mathcal{F}} |\mathcal{A}(F,T)| \geqq (n-2l)(l|\mathcal{F}(T)|) - (n-l). \tag{5.4}$$

Proof of (5.4). In fact, using Claim 5.5 we infer

$$\sum_{T \subset F \in \mathcal{F}} |\mathcal{A}(F,T)| = (n-2l) \sum_{F \in \mathcal{F}(T)} |\{x \in F-T : |\mathcal{F}(T \cup \{x\})| > 1\}|$$

$$= (n-2l) \sum_{x \in X-T, |\mathcal{F}(T \cup \{x\})| > 1} |\mathcal{F}(T \cup \{x\})| \geqslant$$

$$\geqq (n-2l)\left[\sum_{x \in X-T} |\mathcal{F}(T \cup \{x\})| - (n-l) \right] = (n-2l)(l|\mathcal{F}(T)|) - (n-l)). \quad \blacksquare$$

For $A \in \binom{X}{l}$ we have

$$|\{(F,T) : A \in \mathcal{A}(F,T)\}| \leqslant (l+1)l. \tag{5.5}$$

Indeed, for $A \in \mathcal{A}(F,T)$ there exists $y \in A$ such that $A - \{y\}$ is an own subset of F. Thus the number of choices for F is at most l. On the other hand for some $x \in F-A$, $T = F-A-\{x\}$ holds which gives only $l+1$ possibilities. $\quad \blacksquare$

If for some $F, G \in \mathcal{F}$, $F \cup G = X$ holds, then $|\mathcal{F}| = 2$ and (b) follows. Thus we may assume $n \geqslant 2l+1$ and $F \cup G \neq X$ for all $F, G \in \mathcal{F}$.

Next we show:

Claim 5.6. For fixed A : $|\{F \in \mathcal{F} : A \in \mathcal{A}(F,T)\}| \leqslant \min\{l, n-2l\}$ holds.

For $n \geqslant 3l$ we have $l \leqslant n-2l$ thus this assertion follows from the above argument. Assume now $n = 2l+r < 3l$.

Take (F,T) so that $A \in \mathcal{A}(F,T)$. Let us choose $F' \in \mathcal{F}$ satisfying $(F-A) \subset F'$ — this is possible as $|\mathcal{F}(F-A)| > 1$. Since $F \cup F' \neq X$, $|F-F'| \leqslant r-1$.

If $A \in \mathcal{A}(F_0, T_0)$, then for some $y \in A$, $A - y$ is an own subset of F_0. Since $F \not\subseteq F' \cup F_0$, $y \in F - F'$ holds. Thus there are only $r-1$ choices for y and consequently for F_0. Together with F this gives $1 + (r-1) = r = n - 2l$. ∎

Corollary 5.7. For fixed A and T we have

$$|\{F \in \mathcal{F}: A \in \mathcal{A}(F, T)\}| \leqslant \min\{l, n - 2l, |\mathcal{F}(T)|\}. \quad \blacksquare$$

Using Claim 5.6 we obtain the following refinement of (5.5):

$$|\{(F, T) : A \in \mathcal{A}(F, T)\}| \leqslant (l+1) \min\{l, n - 2l\}. \tag{5.6}$$

With these preparations we are ready to define a nonnegative weight function $W_{(F, T)} : \binom{X}{l} \to R^+$ for all $T \subset F \in \mathcal{F}$, $|T| = l$. For convenience we set $l_0 = \min(l, n - 2l)$.

(a) If $|\mathcal{F}(T)| = 1$ then $W_{(F, T)}(A) = \begin{cases} 1 & \text{for } A = T \\ 0 & \text{otherwise} \end{cases}$.

(b) If $1 < |\mathcal{F}(T)| \leqslant (n - l)/l$, then $W_{(F, T)}(A) = l/(n - l)$ for $A = T$ and 0, otherwise.

(c) If $|\mathcal{F}(T)| > (n - l)/l$, then

$$W_{(F, T)}(A) = \begin{cases} 1/|\mathcal{F}(T)|, & \text{for } A = T, \\ 1/(l+1)l_0, & \text{if } A \in A(F', T) \text{ for some } F', \\ 0, & \text{otherwise}. \end{cases}$$

Let us estimate the sum of the weights in the case (c), for brevity we set $|\mathcal{F}(T)| = d$, $\min(l_0, d) = d_0$, and use Corollary 5.7 and (5.4):

$$\sum_{A \in \binom{X}{l}} W_{(F, T)}(A) = \frac{1}{d} + \frac{1}{(l+1)l_0} \left| \left\{ A : A \in \bigcup_{F' \in \mathcal{F}(T)} \mathcal{A}(F', T) \right\} \right|$$

$$\geqslant \frac{1}{d} + \frac{1}{(l+1)l_0} \frac{1}{d_0} \sum_{F' \in \mathcal{F}(T)} |\mathcal{A}(F', T)|$$

$$\geqslant \frac{1}{d} + \frac{(n - 2l)(dl - (n - l))}{l_0(l+1)d_0} \geqslant \frac{1}{d} + \frac{dl - (n - l)}{(l+1)d_0}$$

$$= \frac{l}{n - l} + \frac{(dl - (n - l)(d(n - l) - d_0(l+1))}{(l+1)d_0 d(n - l)} > \frac{l}{n - l}.$$

Using Proposition 5.1 we infer for any fixed $F \in \mathscr{F}$

$$\sum_{T \subset F} \sum_{A \in \binom{X}{l}} W_{(F,T)}(A) \geqslant \frac{1}{2} \binom{2l}{l} + \frac{1}{2} \binom{2l}{l} \frac{l}{n-l} \tag{5.7}$$

$$= \binom{2l-1}{l} \frac{n}{n-l}.$$

On the other hand for any fixed $A \in \binom{X}{l}$ we prove

$$\sum_{(F,T)} W_{(F,T)}(A) \leqslant 1. \tag{5.8}$$

Indeed, if $|\mathscr{F}(A)| \geqslant 1$, then $W_{(F,T)}(A)$ is positive only for $T = A$ and even then it is at most $1/|\mathscr{F}(A)|$; if $|\mathscr{F}(A)| = 0$, then (5.8) follows from (5.6).

Now using inequalities (5.7) and (5.8) we infer:

$$|\mathscr{F}| \binom{2l-1}{l} \frac{n}{n-l} \leqslant \sum_{F \in \mathscr{F}} \sum_{T \in \binom{F}{l}} \left[\sum_{A \in \binom{X}{l}} W_{(F,T)}(A) \right] = \sum_{A \in \binom{X}{l}} \left[\sum_{(F,T)} W_{(F,T)}(A) \right] \leqslant \binom{n}{l}.$$

Comparing the two extreme sides yields:

$$|\mathscr{F}| \leqslant \frac{n-l}{n} \binom{n}{l} / \binom{2l-1}{l} = \binom{n-1}{l} / \binom{2l-1}{l}. \quad \blacksquare$$

Remark 5.8. In [EFF1] it is proved also that in (a) and (b) equality can hold only for the corresponding Steiner-systems. For a more general problem see [EFF2].

Next we consider the same problem for the non-uniform case.

Theorem 5.9. Suppose that $\mathscr{F} \subset 2^X$ with no three members F, G, H satisfying $F \subset G \cup H$. Then we have

$$1.134^n \leqslant \max|\mathscr{F}| < 1.25^n \sqrt{8n}. \tag{5.9}$$

Proof. First we prove the upper bound. We shall make use of the inequalities $\binom{2l}{l} > 2^{2l}/\sqrt{4l}$, $\binom{2l+1}{l+1} > 2^{2l}/\sqrt{4l-2}$ valid for $l \geqslant 1$ and easily proved by induction. The first inequality implies for $n \geqslant 2l$

$$\binom{n-1}{l}\Big/\binom{2l-1}{l} \leqslant \binom{n}{l}\Big/\binom{2l}{l} < \sqrt{2n}\,\binom{n}{l}2^{-2l}. \tag{5.10}$$

The second yields

$$\binom{n}{l}\Big/\binom{2l-1}{l} < \sqrt{2n}\,\binom{n}{l}2^{-2l}. \tag{5.11}$$

By Theorem 5.3 we have $f_{2l} \leqslant \binom{n-1}{l}\big/\binom{2l-1}{l}$ and $f_{2l-1} \leqslant \binom{n}{l}\big/\binom{2l-1}{l}$. Thus using (5.10) and (5.11) we infer

$$|\mathscr{H}| = \sum_{i=0}^{n} f_i < 2\sqrt{2n}\sum_{2l \leqslant n}\binom{n}{l}\left[\frac{1}{4}\right]^l < 2\sqrt{2n}\left[1+\frac{1}{4}\right]^n,$$

yielding (5.9).

To prove the lower bound we apply the probabilistic method. Suppose $1 < k < n$ — we will choose k later to be about $0.26n$.

Let us choose independently and with probability $2m/\binom{n}{k}$ each of the k-subsets of X, the value of m will be fixed later. Let $\mathscr{S}$ denote the obtained random hypergraph. Obviously, the expectation of the number of edges in $\mathscr{S}$ is $E(|\mathscr{S}|) = 2m$. We will need the expression for the number $R(n,k)$ of ordered pairs of k-sets B,C such that for a given k-set A the relation $A \subset B \cup C$ holds:

$$R(n,k) = \sum_{0 \leqslant x \leqslant k}\binom{k}{x}\binom{n-k}{k-x}\sum_{0 \leqslant y \leqslant x}\binom{x}{y}\binom{n-k}{x-y}$$

$$= \sum_{0 \leqslant x \leqslant k}\binom{k}{x}\binom{n-k}{k-x}\binom{n-k+x}{x}$$

$$= \sum_{0 \leqslant x \leqslant k}\binom{k}{x}^2\binom{n-k+x}{k} \leqslant \max_{x \leqslant k}\binom{k}{x}^2\binom{n-k+x}{k}n.$$

Thus for $m < 1/2\sqrt{2}\binom{n}{k}\Big/\sqrt{n\max_{x \leqslant k}\binom{k}{x}^2\binom{n-k+x}{k}}$ we have $R(n,k)4m^2l/\binom{n}{k}^2 < \frac{1}{2}$, yielding that the probability for a given edge $A \in \mathscr{S}$ to be covered by $B \cup C$, $B,C \in \mathscr{S}$ is less than $\frac{1}{2}$.

Hence the expected number of edges to remain in $\mathscr{S}$ after the omission of the covered edges is greater than $2m - \frac{1}{2}2m = m$, and in the remaining hypergraph the conditions are already satisfied.

So we have shown the existence of a desired hypergraph with at least $1/2\sqrt{2}\binom{n}{k}\Big/\sqrt{n\max_{x \leqslant k}\binom{k}{x}^2\binom{n-k+x}{k}}$ edges. All we need is a lower bound on this expression.

The ratio of the terms to be maximized, for consecutive values of x, is $(k-x+1)^2(n-k+x)/(n-2k+x)x^2$. This function is monotone decreasing in x, thus the maximum is taken at the value where this ratio is about 1. We get a quadratic equation in x; the solution of which is $(0 \leqslant x \leqslant k)$

$$x_{\max} \sim \frac{1}{2}(3k-2n+\sqrt{5k^2-8kn+4n^2}).$$

Setting $k=0.26n$ we obtain $x_{\max}=0.1413 \cdots n$. Putting this value back into the expression for m and applying the Stirling formula we see that m can be as large as $(1.1348)^n$. ∎

6. Families in which no Member Contains the Intersection of Two Others

Clearly, if $\mathcal{F} \subset 2^X$ has no three sets, F,G,H with $F \subset G \cup H$, then its complement $\bar{\mathcal{F}}$ has no 3 sets with $F^c \supset G^c \cap H^c$. That is, we come back to the problem of the preceding section.

However, for $n-k$ fixed, $F \subset \left[\begin{matrix} X \\ n-k \end{matrix}\right]$, the bounds of Theorem 5.3 provide only the inequality $|\mathcal{F}| < (2+o(1))^k$. This can be considerably improved and for this purpose it is more convenient to look at the complements.

Fix $k, k \geqslant 1$ and suppose $\mathcal{F} \subset \left[\begin{matrix} X \\ k \end{matrix}\right]$ with no three members F,G,H satisfying $F \supset G \cap H$. This condition implies that in particular $E \cap F \neq G \cap H$ holds whenever $E,F,G,H \in \mathcal{F}$ and $\{E,F\} \neq \{G,H\}$. If a family satisfies this weaker condition, we call it *intersection-free*.

Proposition 6.1. If $\mathcal{F} \subset \left[\begin{matrix} X \\ k \end{matrix}\right]$ contains no three members F, G, H with $F \cap G \subset H$, then

$$|\mathcal{F}| \leqslant 1 + \left(\begin{matrix} k \\ \left\lfloor \frac{k}{2} \right\rfloor \end{matrix}\right)$$

holds.

Proof. Just observe that $\{F \cap G : F \neq G \in \mathcal{F}\}$ is a Sperner family on 2^F. ∎

Remark 6.2. Taking $\mathcal{F} = \left[\begin{matrix} \{1,2,...,k+1\} \\ k \end{matrix}\right]$ for $k=1,2,3$ or the complements of the lines of the projective plane of order 2 for $k=4$ gives equality in Proposition 6.1.

For k large there is a much stronger upper bound.

Theorem 6.3. (Kleitman, Shearer, Sturtevant [KSS]). If $\mathcal{F} \subset \binom{X}{k}$ is intersection-free then

$$|\mathcal{F}| \leqslant (1.76 + o(1))^k$$

holds.

For the proof of this theorem a little information theory is needed. Suppose ξ is a discrete random variable taking distinct values from some (countable) set with probabilities $p_1, p_2, \ldots$, then its entropy $H(\xi)$ is defined to be $H(\xi) = -\sum_{p_i > 0} p_i \log p_i$.

Suppose now that $\vec{\xi} = (\xi_1, \ldots, \xi_n)$ is vector-valued with ξ_i taking values from S_i. The ξ_i's are called the coordinates of $\vec{\xi}$. For $0 < \alpha < 1$ we set $H(\alpha) = -\alpha \log \alpha - (1-\alpha) \log (1-\alpha)$, $H(0) = H(1) = 0$.

Proposition 6.4. (*cf.* McEliece [Mc]).

$$H(\vec{\xi}) \leqslant \sum_i H(\xi_i). \tag{6.1}$$

We use (6.1) to prove

Proposition 6.5. (Kleitman, Shearer, Sturtevant [KSS]). Suppose $\mathcal{F} \subset 2^X$. Then we have

$$\log |\mathcal{F}| \leqslant \sum_{1 \leqslant i \leqslant n} H\left(\frac{|\mathcal{F}(i)|}{|\mathcal{F}|} \right). \tag{6.2}$$

Proof. Let $\vec{\xi}$ be the uniformly distributed random variable defined by $p(\vec{\xi} = F) = \dfrac{1}{|\mathcal{F}|}$ for $F \in \mathcal{F}$. By definition $H(\vec{\xi}) = \log |\mathcal{F}|$.

Now $\vec{\xi}$ can be considered as a vector-valued distribution: $\vec{\xi} = (\xi_1, \ldots, \xi_n)$ where $\xi_i = \begin{cases} 1 & \text{if } i \in \vec{\xi} \\ 0 & \text{if } i \notin \vec{\xi} \end{cases}$

Then $H(\xi_i) = H(|\mathcal{F}(i)|/|\mathcal{F}|)$. Thus (6.1) implies (6.2). ∎

Proof of Theorem 6.3. Let us set $\mathcal{G} = \{F \cap F' : F, F' \in \mathcal{F}\}$. Since $\mathcal{F}$ is intersection-free $|\mathcal{G}| = \binom{m+1}{2}$.

By definition $|\mathcal{G}(i)| = \binom{|\mathcal{F}(i)| + 1}{2}$, i.e.,

$$|\mathscr{G}(i)|/|\mathscr{G}| = (|\mathscr{F}(i)|/|\mathscr{F}|)^2 + o(1).$$

For convenience set $\delta_i = |\mathscr{F}(i)|/|\mathscr{F}|$. Applying (6.2) gives

$$\log \binom{m+1}{2} \leqslant \sum_{1 \leqslant i \leqslant n} H(\delta_i^2)(1+o(1)) = k(1+o(1)) \sum_i \frac{\delta_i}{k}\,\frac{H(\delta_i^2)}{\delta_i}. \tag{6.3}$$

To estimate the RHS of (6.3) note that

$$\sum_i \frac{\delta_i}{k} = \frac{1}{k|\mathscr{F}|}\sum_i |\mathscr{F}(i)| = 1, \quad \text{since} \quad \mathscr{F} \subset \binom{X}{k}.$$

Also, the function $H(x^2)/x$ is concave for $0 \leqslant x \leqslant 1$. Thus the Jensen inequality gives with

$$\delta = \sum \frac{\delta_i}{k}\,\delta_i$$

$$\sum \frac{\delta_i}{k}\,\frac{H(\delta_i^2)}{\delta_i} \leqslant H(\delta^2)/\delta.$$

Now substituting into (6.3) we obtain

$$\log m < \frac{k}{2}(1+o(1)) \max_{0 \leqslant \delta \leqslant 1} H(\delta^2)/\delta.$$

Easy computation shows that this maximum is attained for $\delta \sim 0.4914$ where $H(\delta^2)/\delta \sim 1.6228$. Consequently $m \leqslant 2^{k(0.8114+o(1))}$ holds, proving Theorem 6.3. $\blacksquare$

Remark 6.6. The proof of the lower bound of Theorem 5.9 can be applied to show the existence of a k-uniform $\mathscr{F}$ without 3 sets F,G,H satisfying $F \supset G \cap H$ and with $|\mathscr{F}| \geqslant (1.2+o(1))^k$.

Let us note, that for this second problem Kleitman et al. [KSS] obtained an upper bound better than the one in Theorem 6.3.

7. Union-Free Families: Uniform Case

A family $\mathscr{F}$ is called *union-free* if all the $\binom{|\mathscr{F}|}{2}$ pairwise unions $F \cup F'$ with $F,F' \in \mathscr{F}$ are distinct. There are two natural weakenings of this property. If we require $F \cup F' \neq G \cup G'$ only for any 4 distinct members of $\mathscr{F}$, then $\mathscr{F}$ is called *weakly union-free*.

If $\mathscr{F}$ satisfies $F \cup G \neq F \cup H$ for all $F,G,H \in \mathscr{F}$ then $\mathscr{F}$ is called *cancellative*. Note that $\mathscr{F}$ is union-free if and only if it is both weakly union-free and cancellative.

Let us first illustrate this terminology by considering the case of graph is, i.e., $\mathscr{F} \subset \binom{X}{2}$. Then $\mathscr{F}$ is weakly union-free iff it does not contain C_4 (i.e., 4 edges of the form (x,y), (y,z), (z,u), (u,x)). It is cancellative iff it does not contain C_3, a triangle. The last case is easier. By a theorem of Mantel [M], from the beginning of this century, we know that if $\mathscr{F}$ has no triangle it must satisfy $|\mathscr{F}| \leqslant \left\lfloor \dfrac{n^2}{4} \right\rfloor$, equality holds iff $\mathscr{F}$ is $K_{\left\lfloor \frac{n}{2} \right\rfloor, \left\lfloor \frac{n}{2} \right\rfloor}$, the complete bipartite graph.

Let us denote by $u_k(n)$ $(w_k(n), c_k(n))$ the maximum size of a union-free (weakly union-free, cancellative) family $F \subset \binom{X}{k}$, respectively.

The problem of determining the maximum number of edges in a graph without C_4 goes back to Erdös [E1], 1938. However, the exact answer is known only for $n = p^{2\alpha} + p^{\alpha} + 1$, p a prime, $\alpha \geqslant 1$.

The construction, due to Erdös and Rényi [ER], is as follows. Let Γ be the finite field of p^{α} elements and consider the projective plane over Γ, that is, all 3-tuples (x_1, x_2, x_3) not all zero, $x_i \in \Gamma$. Identify (x_1, x_2, x_3) and (ax_1, ax_2, ax_3) for all $a \in \Gamma$, $a \neq 0$. These equivalence classes (the points of the plane) are the vertices. Put an edge between two of them iff $x_1 y_1 + x_2 y_2 + x_3 y_3 = 0$. Recently, Füredi [Fü1] showed, that for $n = p^{2\alpha} + p^{\alpha} + 1$ the Erdös-Rényi graph has the maximum possible of edges among all graphs without C_4.

Since primes are sufficiently densely distributed, e.g., the ratio of consecutive primes tends to 1, we have $w_2(n) = \left\lfloor \dfrac{1}{2} + o(1) \right\rfloor n^{3/2}$ (cf. Kővári, Sós, Turán [KST]).

For union-free graphs the situation is worse. The best known bounds are

$$\frac{1 + o(1)}{2\sqrt{2}} \, n^{3/2} \leqslant u_2(n) \leqslant w_2(n) = \left\lfloor \frac{1}{2} + o(1) \right\rfloor n^{3/2}.$$

The lower bound follows again by a construction using projective planes: let $P = \{p_1, ..., p_m\}$ be the points, $L = \{l_1, ..., l_m\}$ the lines of a finite projective plane. Make a graph on $P \cup L$ by joining each l by an edge to all p's which lie on l.

Being bipartite, this graph has no triangle. Since any two lines intersect in only one point, it cannot contain $C_4 = K_{2,2}$ either.

Suppose $X = X_1 \oplus X_2 \oplus \cdots \oplus X_k$ and $\mathscr{F}$ is the complete k-partite graph, i.e., $\mathscr{F} = \{F \in \binom{X}{k} : |F \cap X_i| = 1\}$. Then $\mathscr{F}$ is cancellative and $|\mathscr{F}| = |X_1| \cdots |X_k|$. Clearly this product is maximal iff the $|X_i|$'s are as equal as possible, e.g.,

$$|X_i| = \left\lfloor \frac{n+i-1}{k} \right\rfloor.$$

Theorem 7.1. (Bollobás [B]). $c_3(n) = \left\lfloor \dfrac{n}{3} \right\rfloor \left\lfloor \dfrac{n+1}{3} \right\rfloor \left\lfloor \dfrac{n+2}{3} \right\rfloor$ and $|\mathscr{F}| = c_3(n)$ holds for a cancellative family only if it is complete 3-partite.

Proof. Suppose we are given a cancellative family, $\mathscr{F} \subset \binom{X}{3}$. Let us select $P \in \binom{X}{2}$ so that $|\mathscr{F}(P)|$ is maximal. Set $A = \mathscr{F}(P)$. Select next Q so that $|\mathscr{F}(Q)|$ is maximal over all $Q \in \binom{X}{2}$, $Q \cap A \neq \varnothing$. Set $B = \mathscr{F}(Q)$. Finally select R so that $\mathscr{F}(R)$ is maximal subject to $|R| = 2$, $|R \cap A| = |R \cap B| = 1$. Set $C = \mathscr{F}(R)$, $D = X - (A \cup B \cup C)$. Define $|A| = a$, $|B| = b$, $|C| = c$, $|D| = d$.

Proposition 7.2.

(i) $X = A \oplus B \oplus C \oplus D$,

(ii) no $F \in \mathscr{F}$ intersects A, B or C in more than 1 element.

Proof of Proposition 7.2. (i) Suppose that there exists $z \in A \cap B$. Since $Q \cap A \neq \varnothing$, we may choose $w \in Q \cap A$. Then $F = P \cup \{z\}$, $G = P \cup \{w\}$ and $H = Q \cup \{z\}$ are all in $\mathscr{F}$, however $F \cup H = G \cup H$, a contradiction. $A \cap C = \varnothing$ and $B \cap C = \varnothing$ can be proved in the same way.

(ii) We claim that for every $T \in \binom{X}{2}$ and any $F \in \mathscr{F}$, $|F \cap \mathscr{F}(T)| \leq 1$ holds. In fact, otherwise choose x, y from the intersection and note $F \cup (T \cup \{x\}) = F \cup (T \cup \{y\})$, a contradiction. To prove (ii) set successively $T = P, Q, R$. ∎

Since $\displaystyle\sum_{F \in \mathscr{F}} \left| \binom{F}{2} \right| = 3|\mathscr{F}|$ and a 2-element set T can be contained in some member of $\mathscr{F}$ only in one of the following three cases: a) $|T \cap A| = 1 = |T \cap B|$, b) $|T \cap A| = 1$, $T \cap B = \varnothing$ c) $T \cap A = \varnothing$, Proposition 7.2 implies

$$3|\mathscr{F}| \leqslant abc + a(n-a-b)b + \left[\binom{n-a}{2} - \binom{b}{2} - \binom{c}{2}\right]a \overset{\text{def}}{=} f(a,b,c) \qquad (7.1)$$

Clearly, equality in (7.1) can hold only if $a+b+c = n$ (otherwise we get a contradiction to the definition of c).

If $n = a+b+c$, then $\mathscr{F}$ is a sub-system of the complete 3-partite graph, thus we have nothing to prove. Thus it is sufficient to show that $d > 0$ implies that the RHS of (7.1) is less than

$$\left\lfloor \frac{n}{3} \right\rfloor \left\lfloor \frac{n+1}{3} \right\rfloor \left\lfloor \frac{n+2}{3} \right\rfloor.$$

This is now pure arithmetic: Choose the integers $a, b \geqslant c \geqslant 0$ so as to maximize (7.1) subject to $a+b+c \leqslant n$. Suppose for contradiction $a+b+c \leqslant n-1$. Since $\dfrac{\partial f(a,b,c)}{\partial c} = a\left(b - c + \dfrac{1}{2}\right)$ is positive for $b > c$, in that case $f(a,b,c+1) > f(a,b,c)$, a contradiction. Thus we may assume $b = c$. Also $\dfrac{\partial f(a,b,c)}{\partial b} = a\left(n-a-b-c - 2(b-c) + \dfrac{1}{2}\right)$, which implies $f(a,b+1,c) > f(a,b,c)$, for $n-a-b-c \geqslant 1$. $\blacksquare$

For (weakly) union-free families we have

Theorem 7.3. (Frankl, Füredi [FF1]). (i) $u_3(n) = \left\lfloor \dfrac{n(n-1)}{6} \right\rfloor$, moreover any Steiner triple-system provides equality.

(ii) $w_3(n) \leqslant \dfrac{n(n-1)}{3}$, if $\mathscr{F}$ attains equality, then $\mathscr{F}$ is necessarily a $S_2(n,3,2)$ (*cf.* Definition 7.5).

Proof. Suppose $\mathscr{F} \subset \binom{X}{3}$. Define for $0 \leqslant i \leqslant n-2$

$$\mathscr{G}_i = \left\{P \in \binom{X}{2} : |\mathscr{F}(P)| = i\right\}, \quad g_i = |\mathscr{G}_i|.$$

The next two equalities are obvious.

$$\sum_i g_i = \binom{n}{2} \tag{7.2}$$

$$\sum_i i\, g_i = 3|\mathscr{F}|. \tag{7.3}$$

Proposition 7.4.

a) if $\mathscr{F}$ is weakly union-free then $|\mathscr{F}(P) \cap \mathscr{F}(Q)| \leq 2$ holds for all $P, Q \in \binom{X}{2}$.

b) if $\mathscr{F}$ is union-free then $\binom{\mathscr{F}(P)}{2} \subset \mathscr{G}_0$ holds for all $P \in \binom{X}{2}$.

Proof. Suppose for contradiction $x, y \in |\mathscr{F}(P) \cap \mathscr{F}(Q)|$. Then $P \cup \{x\}, Q \cup \{y\}, Q \cup \{x\}$, $P \cup \{y\} \in \mathscr{F}$. However, the union of the first two equals that of the last two, a contradiction, proving a).

To prove b) suppose $x, y \in \mathscr{F}(P)$ and $\{x, y\} \notin \mathscr{G}_0$. That is $\{x, y\} \subset F \in \mathscr{F}$ holds for some $F \in \mathscr{F}$. However, $F \cup (P \cup \{x\}) = F \cup (P \cup \{y\})$, a contradiction. $\blacksquare$

Clearly, Proposition 7.4 implies

$$\sum_{i \geq 1} \binom{i}{2} g_i \leq \binom{n}{2} \quad \text{for weakly union-free families and} \tag{7.4}$$

$$\sum_{i \geq 1} \binom{i}{2} g_i \leq g_0 \quad \text{for union-free families.} \tag{7.5}$$

Suppose $\mathscr{F}$ is union-free. Using (7.5) and (7.2) we obtain

$$\binom{n}{2} = g_0 + \sum_{i \geq 1} g_i \geq \sum_{i \geq 1} \left[1 + \binom{i}{2}\right] g_i.$$

Since $1 + \binom{i}{2} \geq i$ holds for $i \geq 1$, (7.3) implies $\binom{n}{2} \geq \sum_{i \geq 1} i g_i = 3|\mathscr{F}|$, i.e., $|\mathscr{F}| \leq \left\lfloor \dfrac{n(n-1)}{6} \right\rfloor$, the

desired upper bound.

If $\mathscr{F}$ is weakly union-free, then summing up (7.2) and (7.4), (7.3) implies:

$$2\binom{n}{2} \geq \sum_{i \geq 0} \left[1 + \binom{i}{2}\right] g_i \geq \sum i g_i = 3|\mathscr{F}|;$$

that is $|\mathscr{F}| \leq \dfrac{n(n-1)}{3}$, as desired. If one has equality, $g_i = 0$ must hold whenever $1 + \binom{i}{2} > i$, i.e.,

$g_i \neq 0$ can hold only for $i = 1, 2$. Thus $g_1 = \binom{n}{2} - g_2$. Putting this back into (7.3) gives: $\binom{n}{2} + g_2 = n(n-1)$, i.e., $g_2 = \binom{n}{2}$, $g_1 = \binom{n}{2} - g_2 = 0$. In words, every two-subset of X is contained in exactly two members of $\mathscr{F}$ — such a system is called a $S_2(n, 3, 2)$.

Definition 7.5. A family $\mathscr{F} \subset \binom{X}{k}$ is called a $S_\lambda(n, k, t)$ if every $T \in \binom{X}{t}$ is contained in exactly λ members of $\mathscr{F}$.

The reader can easily check that any $S_1(n, 3, 2)$ provides a union-free family with $|\mathscr{F}| = \dfrac{n(n-1)}{6}$. It is known that $S_1(n, 3, 2)$ exists iff $n \equiv 1$ or $3 \pmod 6$ while $S_2(n, 3, 2)$ exists iff $n \equiv 0, 1 \pmod 3$.

The proof of the lower bound for $u_3(n)$ in the cases $n \not\equiv 1$ or $3 \pmod 6$ is left for the reader. ■

For weakly union-free families we have the following example.

Suppose $n \equiv 1 \pmod 3$, n is a prime power and Γ is the field of n elements. Choose $\gamma \in \Gamma$, $\gamma \neq 1$ with $\gamma^3 = 1$, it is possible because $3 \mid (n-1)$.

Example 7.6. $\mathscr{F} = \left\{ \{a, b, c\} : a + b\gamma + c\gamma^2 = 0 \right\}$.

Claim 7.7. $\mathscr{F}$ is a $S_2(n, 3, 2)$.

Proof. To given $\{a, b\} \in \binom{\Gamma}{2}$, $\{a, b, c\} \in \mathscr{F}$ if and only if $a + b\gamma + c\gamma^2 = 0$ or $a\gamma + b + c\gamma^2 = 0$. Equivalently $c = -a\gamma - b\gamma^2$ or $c = -a\gamma^2 - b\gamma$. ■

Proposition 7.8. $\mathscr{F}$ is weakly union-free.

Proof. Suppose for contradiction $F \cup F' = G \cup G'$ holds for 4 distinct members of $\mathscr{F}$. We want to derive a contradiction. We distinguish 3 cases:

(i) $|F \cup F'| = 6$.

We may write $F = \{a, b, c\}$, $F' = \{x, y, z\}$, $G = \{a, b, z\}$, $G' = \{x, y, c\}$. Exchanging, if necessary, x and y or a and b we may suppose

$$c = -a\gamma - b\gamma^2 \qquad\qquad z = -a\gamma^2 - b\gamma$$

$$c = -x\gamma^2 - y\gamma \qquad\qquad z = -x\gamma - y\gamma^2$$

Adding up the equalities in one line and comparing gives $c + z = -(a+b)(\gamma + \gamma^2) = -(x+y)(\gamma + \gamma^2)$.

Thus $a + b = x + y$. Substituting this into the first equalities in both rows:

$$c = -(a+b)\gamma + b(\gamma - \gamma^2) = -(x+y)\gamma + b(\gamma - \gamma^2)$$

$$c = -(x+y)\gamma + y(\gamma - \gamma^2).$$

Comparing the RHS's yields $b = y$, contradicting $|F \cup F'| = 6$.

(ii) $|F \cup F'| = 5$, i.e., $F = \{a,b,c\}$, $F' = \{a,y,z\}$, $G = \{a,b,z\}$, $G' = \{u,y,c\}$ where $u \in \{a,b,z\}$.

If $u = a$, then $y = b$ follows as in case (i). That is $|F \cup F'| \leqslant 4$, a contradiction.

If $u = b$ then we have:

$$c = -a\gamma - b\gamma^2, \quad z = -a\gamma^2 - b\gamma$$

$$c = -y\gamma^2 - b\gamma, \quad z = -a\gamma - y\gamma^2.$$

Which lead to

$$(a-b)(\gamma^2 - \gamma) = c - z = (a-b)\gamma,$$

that is $(a-b)(\gamma^2 - 2\gamma) = 0$, i.e., $a = b$, a contradiction.

The case $u = z$ is equivalent to the preceding one after renaming the variables.

(iii) $|F \cup F'| = 4$, i.e., $\{F,F',G,G'\} = \binom{\{a,b,c,d\}}{3}$, that is both $c = -a\gamma - b\gamma^2$, $d = -a\gamma^2 - b\gamma$ and one of

$$a = -c\gamma - d\gamma^2 \quad \text{or} \quad a = -c\gamma^2 - d\gamma \quad \text{hold.}$$

In both cases we express c and d by a and b using the first two equations. If $a = -c\gamma - d\gamma^2$, we infer $a = a\gamma^2 + b + a\gamma + b$. Using $1 + \gamma + \gamma^2 = 0$, this leads to $2a = 2b$, i.e., $a = b$, a contradiction. Similarly $a = -c\gamma^2 - d\gamma$ leads to $a = b$. ∎

Let us note that not all $S_2(n,3,2)$ are weakly union-free. In fact, Colbourn and Rosa showed that no $S_2(n,3,2)$ with $n < 10$ is weakly union-free.

For $k \geqslant 4$ there are only upper and lower bounds for $c_k(u)$, $u_k(u)$, $w_k(n)$.

Conjecture 7.9. (Bollobás [B]). $c_k(u) = \prod_{i=0}^{k-1} \left\lfloor \dfrac{n+i}{k} \right\rfloor$ where equality holds only if $\mathcal{F}$ is a k-partite complete graph.

Let us mention that recently Sidorenko [Si] proved that Conjecture 7.9 is true for $k = 4$.

Proposition 7.10. (Frankl, Füredi [FF3]). For $k \leqslant n \leqslant 2k$

$$c_k(n) = 2^{n-k}. \tag{7.4}$$

Proof. Suppose $\mathcal{F} \subset \binom{X}{k}$, $\mathcal{F}$ is cancellative. Fix $F_0 \in \mathcal{F}$. Then clearly $F \cap (X - F_0) \neq F' \cap (X - F_0)$ for all $F, F' \in \mathcal{F}$. Thus

$$|\mathcal{F}| \leqslant 2^{|X - F_0|} = 2^{n-k}. \quad \blacksquare$$

Let us introduce a new notion, call $\mathcal{F}$ *intersection or union-free* if for any four distinct $A, B, A', B' \in \mathcal{F}$ either $A \cup B \neq A' \cup B'$ or $A \cap B \neq A' \cap B'$ holds. We denote by $v_k(n)$ the maximum size of an intersection or union-free family.

Proposition 7.11.

$$v_k(n) \geqslant w_k(n) \geqslant u_k(n) \geqslant \frac{k!}{k^k} v_k(n).$$

Proof. The first two inequalities are trivial, to prove the last, take an intersection or union-free family $\mathcal{F} \subset \binom{X}{k}$, with $|\mathcal{F}| = v_k(n)$. By Theorem 4 $\mathcal{F}$ contains a k-partite subfamily $\mathcal{F}_0$ with $|\mathcal{F}_0| \geqslant k!/k^k |\mathcal{F}|$. The statement will follow if we show that F_0 is union-free. Suppose $X = X_1 \oplus X_2 \oplus \cdots \oplus X_k$ and $\mathcal{F}_0$ is k-partite with respect to this partition. Assume for contradiction that $A, B, A', B' \in \mathcal{F}_0$ and $A \cup B = A' \cup B'$. Then clearly $A \cap B \cap X_i \neq \varnothing$ if and only if $A \cap X_i = B \cap X_i$ which implies $A \cap X_i = A' \cap X_i = B' \cap X_i$. Thus $A \cap B = A' \cap B'$ holds also. Since $\mathcal{F}_0$ is intersection or union-free, $\{A, B\} = \{A', B'\}$. $\quad \blacksquare$

Theorem 7.12. (Frankl, Füredi [FF2]).

$$\left[\frac{1}{k!}-o(1)\right]n^{2t+1} \leqslant v_{3t+1}(n) \leqslant \left[\frac{t!}{k!}+o(1)\right]n^{2t+1} \quad \text{if } k=3t+1. \tag{7.5}$$

$$\left[\frac{1}{k!}-o(1)\right]n^{2t} \leqslant v_{3t}(n) \leqslant \frac{\sqrt{\left[\begin{smallmatrix}2t\\t\end{smallmatrix}\right]}+1+o(1)}{\left[\begin{smallmatrix}3t\\t\end{smallmatrix}\right](2t)!} n^{2t} \quad \text{if } k=3t. \tag{7.6}$$

$$\left[\frac{1}{(2k)!}-o(1)\right]n^{(4t+3)/2}2^{(4t+3)/2} \leqslant v_{3t+2}(n) \leqslant$$

$$\frac{\sqrt{\left[\begin{smallmatrix}2t+1\\t+1\end{smallmatrix}\right]/(t+1)}+o(1)}{\left[\begin{smallmatrix}3t+2\\2t+1\end{smallmatrix}\right](2t+1)!} n^{(4t+3)/2} \quad \text{if } k=3t+2.$$

Let us note that this theorem together with Proposition 7.11 implies

Corollary 7.13. There exist positive constants a_k, b_k so that

$$a_k n^{\left\lceil\frac{4k}{3}\right\rceil/2} \leqslant u_k(n) \leqslant w_k(n) \leqslant b_k n^{\left\lceil\frac{4k}{3}\right\rceil/2}.$$

Proof of the upper bound part of Theorem 7.12.

Let us set $s=\left\lceil\frac{2k-1}{3}\right\rceil$. Suppose $\mathscr{F}\subset\binom{X}{k}$, $\mathscr{F}$ contains no four distinct sets A,B,A',B' satisfying $A\cup B=A'\cup B'$, $A\cap B=A'\cap B'$. For $S\in\binom{X}{s}$ let us set $\mathscr{F}(S)=\{F-S : S\subset F\in\mathscr{F}\}$ and $d_{\mathscr{F}}(S)=|\mathscr{F}(S)|$. Obviously:

$$\sum_{S\in\binom{X}{s}} d_{\mathscr{F}}(S) = \sum_{F\in\mathscr{F}} \binom{|F|}{s} = \binom{k}{s}|\mathscr{F}|. \tag{7.8}$$

Let us set $a=\dfrac{|\mathscr{F}|\binom{k}{s}}{\binom{n}{s}}$. Using the inequality between the arithmetic and quadratic means gives:

$$\sum_{S\in\binom{X}{s}} (d_{\mathscr{F}}(S))^2 \geqslant \binom{n}{s}a^2.$$

Combining this with (7.8) we obtain

$$\sum_{S\in\binom{X}{s}} \binom{d_{\mathscr{F}}(S)}{2} \geqslant \binom{n}{s}\binom{a}{2}. \tag{7.9}$$

On the other hand for $S, S' \in \binom{X}{s}$, $T, T' \in (\mathscr{F}(S) \cap \mathscr{F}(S'))$ implies that the four distinct sets

$$A = S \cup T, \qquad B = S' \cup T', \qquad A' = S \cup T', \qquad B' = S' \cup T \qquad \text{satisfy} \qquad A \cup B = A' \cup B',$$

$A \cap B = (S \cap S') \cup (T \cap T') = A' \cap B'$. Consequently the sets $\binom{\mathscr{F}(S)}{2}$ are pairwise disjoint for

$S \in \binom{X}{s}$, yielding

$$\sum_{S \in \binom{X}{s}} \binom{d_{\mathscr{F}}(S)}{2} \leqslant \binom{\binom{n}{k-s}}{2}. \tag{7.10}$$

Combining (7.9) and (7.10) we obtain

$$\binom{n}{s} a\,(a-1) \leqslant \binom{n}{k-s}^2. \tag{7.11}$$

If $k = 3t+1$ then $s = 2t+1$ and (7.11) yields

$$|\mathscr{F}| \leqslant (1+o(1)) \binom{n}{2t+1} \Big/ \binom{k}{2t+1},$$

as desired.

If $k = 3t$ then $s = 2t$. Now (7.11) yields $a\,(a-1) \leqslant \binom{2t}{t}$ and consequently $a < \sqrt{\binom{2t}{t}+1}$, that

is, $|\mathscr{F}| < \binom{n}{2t} \left[\sqrt{\binom{2t}{t}} + 1 \right] \Big/ \binom{3t}{t}$, proving (7.6).

If $k = 3t+2$ then $s = 2t+1$. Now (7.11) yields

$$a\,(a-1) \leqslant (1+o(1)) \frac{n}{t+1} \binom{2t+1}{t+1},$$

consequently

$$|\mathscr{F}| \leqslant n^{\frac{4t+3}{2}} \frac{\sqrt{\binom{2t+1}{t+1}/(t+1)}}{\binom{3t+2}{2t+1}(2t+1)!} (1+o(1)). \qquad \blacksquare$$

The proof of the lower bounds in Theorem 7.12. We will first prove (7.5) and then show that this implies (7.6) and (7.7). We need a result of algebraic geometric character. To state it, we need some definitions.

Let K be any field (not necessarily finite). For any subset $Y = \{y_1,...,y_f\}$ of K and any integer i, $i \geq 0$ we denote by $\sigma_i(Y)$ the i'th elementary symmetric polynomial in the variables $y_1,...,y_f$, that is

$$\sigma_i(Y) = \sum_{I \in \left[\genfrac{}{}{0pt}{}{\{1,...,f\}}{i}\right]} \prod_{\nu \in I} y_\nu.$$

In particular $\sigma_0(Y) = 1$, also $\sigma_i(Y) = 0$ for $i > f$ and $i < 0$. For any fixed Y and l let us define an l by l matrix $D_l(Y)$. Let the general entry of $D_l(Y)$ be d_{ij} where $d_{ij} = \sigma_{2i-j}(Y)$.

Theorem 7.14. Suppose $k = 3t+1$, $t \geq 1$, $c_2, c_4,...,c_{2t}$ are arbitrary but fixed elements of K and $\mathscr{F} \subset \left[\genfrac{}{}{0pt}{}{K}{k}\right]$ consists of those k-tuples $A = \{x_1,...,x_k\}$ for which

$$\sigma_{2i}(A) = c_{2i} \quad \text{holds} \quad i = 1,...,t.$$

Moreover, for every subset $Y \subset A$ we have

$$\det D_l(Y) \neq 0, \quad l = 1,...,|Y|-1.$$

Then $\mathscr{F}$ contains no four distinct sets A, B, A', B' satisfying

$$A \cup B = A' \cup B', \quad A \cap B = A' \cap B'.$$

This theorem and the next 3 propositions give the lower bound part of (7.5).

Proposition 7.15. Suppose $|Y| > l$. Then the polynomial $\det D_l(Y)$ is not the zero polynomial.

Proposition 7.16. Suppose $|K| = q$, in particular K is finite. Then the number of k-tuples $A \in \left[\genfrac{}{}{0pt}{}{K}{k}\right]$ for which $\det D_l(Y) = 0$ holds for some l and some $Y \subset A$, $l < |Y|$ is bounded from above by $2^k k^2 \left[\genfrac{}{}{0pt}{}{q}{k-1}\right]$.

Proposition 7.17. There exist constants $c_2, c_4,...,c_{2t} \in K$ such that the family $\mathscr{F}$ defined in Theorem 7.14 satisfies

$$|\mathscr{F}| \geq \frac{\left[\genfrac{}{}{0pt}{}{q}{k}\right]}{q^t} - 2^k k^2 \left[\genfrac{}{}{0pt}{}{q}{k-1}\right] / q^t.$$

In fact, to derive the lower bound of (7.5) choose the largest prime power q, satisfying $q \leq n$, $X \supset K$ and the family $\mathscr{F} \subset \left[\genfrac{}{}{0pt}{}{X}{k}\right]$ from Theorem 7.14. ∎

The proof of Theorem 7.14. Suppose that for some A, A', B, B' satisfying the assumptions $A \cup B = A' \cup B'$ and $A \cap B = A' \cap B'$ hold. We have to prove $\{A,B\} = \{A',B'\}$. Let us define $C = A \cap A'$, $D = A \cap (B'-A')$, $C' = B \cap B'$, $D' = B \cap (A'-B')$. Then these four sets satisfy $C \cap D = C \cap D' = C' \cap D = C' \cap D' = \varnothing$, $C \cup D = A$, $C \cup D' = A'$, $C' \cup D = B'$, $C' \cup D' = B$. Consequently, we have:

$$\sigma_{2i}(C \cup D) = \sigma_{2i}(C \cup D') = \sigma_{2i}(C' \cup D) = \sigma_{2i}(C' \cup D') = c_{2i}$$

for $0 \leqslant i \leqslant t$.

Proposition 7.18. There exist non-negative integers a,b such that $a+b=t$ and $\sigma_i(C) = \sigma_i(C')$, $\sigma_j(D) = \sigma_j(D')$ hold for $0 \leqslant i \leqslant 2a$, $0 \leqslant j \leqslant 2b$.

Proof. Let a and b be the largest integers for which $\sigma_i(C) = \sigma_i(C')$ and $\sigma_j(D) = \sigma_j(D')$ hold for $0 \leqslant i \leqslant 2a$, $0 \leqslant j \leqslant 2b$, respectively. Assume $a+b < t$. Let us write out the equation for $\sigma_{2a+2b+2}$:

$$\sigma_{2a+2b+2}(C \cup D) = \sum_{0 \leqslant i \leqslant 2a+2b+2} \sigma_i(C)\sigma_{2a+2b+2-i}(D) = c_{2a+2b+2}.$$

Denoting this equation by e_{11} and the analogous equations for $C' \cup D'$, $C \cup D'$, $C' \cup D$ by e_{22}, e_{12}, e_{21}, respectively, $e_{11} + e_{22} - e_{12} - e_{21}$ reads:

$$\sum_{0 \leqslant i \leqslant 2a+2b+2} (\sigma_i(C) - \sigma_i(C'))(\sigma_{2a+2b+2-i}(D) - \sigma_{2a+2b+2-i}(D')) = 0.$$

Since $\sigma_i(C) = \sigma_i(C')$ for $0 \leqslant i \leqslant 2a$ and $\sigma_j(D) = \sigma_j(D')$ for $0 \leqslant j \leqslant 2b$, this equation reduces to:

$$(\sigma_{2a+1}(C) - \sigma_{2a+1}(C'))(\sigma_{2b+1}(D) - \sigma_{2b+1}(D')) = 0.$$

Assume by symmetry $\sigma_{2a+1}(C) = \sigma_{2a+1}(C')$. Write now the equations involving σ_{2a+2}:

$$\sigma_{2a+2}(C \cup D) = \sum_{0 \leqslant i \leqslant 2a+2} \sigma_i(C)\sigma_{2a+2-i}(D) = c_{2a+2}.$$

Subtracting from this the corresponding equation for C' and D we obtain ($\sigma_0(D) = 1$):

$$\sigma_{2a+2}(C) - \sigma_{2a+2}(C') + \sum_{0 \leqslant i \leqslant 2a+1} (\sigma_i(C) - \sigma_i(C'))\sigma_{2a+2-i}(D)$$

$$= \sigma_{2a+2}(C \cup D) - \sigma_{2a+2}(C' \cup D) = 0.$$

Since $\sigma_i(C) = \sigma_i(C')$ for $0 \leqslant i \leqslant 2a+1$, $\sigma_{2a+2}(C) = \sigma_{2a+2}(C')$ follows, contradicting the maximal choice of a. ∎

If $|C| \leqslant 2a$, then $C = C'$ and thus $A = B'$, $A' = B$ i.e., $\{A,B\} = \{A',B'\}$ follows. Similarly, in the case $|D| \leqslant 2b$. Thus we may assume $|C| > 2a$, $|D| > 2b$.

Lemma 7.19. Either $|C| \leqslant t+a$ or $|D| \leqslant t+b$ holds.

Proof. Suppose the contrary. Then we infer — using $|C| + |D| = 3t+1$:

$$3t+1 \geqslant 2t+a+b+2 \quad \text{or equivalently} \quad t \geqslant a+b+1,$$

contradicting the choice of a,b. ∎

Assume, by symmetry: $0 < |D| \leqslant t+b$ holds.

Lemma 7.20. $\sigma_j(D) = \sigma_j(D')$ holds for all $j \geqslant 0$.

Note that this lemma yields $D = D'$ and consequently $\{A,B\} = \{A',B'\}$, i.e., it concludes the proof of Theorem 7.14.

Proof of the Lemma. For $i = 1,...,t-b$ let e_i (e'_i) denote the equation $\sigma_{2b+2i}(C \cup D) = c_{2b+2i}$ $(\sigma_{2b+2i}(C \cup D') = c_{2b+2i})$, respectively. Now $e_i - e_i$ reads (using $\sigma_j(D) = \sigma_j(D')$ for $0 \leqslant j \leqslant 2b$):

$$\sum_{0 \leqslant j < 2i} \sigma_j(C)(\sigma_{2b+2i-j}(D) - \sigma_{2b+2i-j}(D')) = 0.$$

This is a homogeneous system of linear equations in the unknowns $\sigma_{2b+\nu}(D) - \sigma_{2b+\nu}(D')$ for $\nu = 1,...,t+b$, because $|D| \leqslant t-b$ by our assumptions, i.e., $c_j(D) = 0$ for $j > t+b$. By the assumptions of Theorem 7.14 the determinant of this system, $\det D_{t-b}(C)$ is nonzero, consequently $\sigma_{2b+\nu}(D) = \sigma_{2b+\nu}(D')$ holds for $\nu = 1,...,t-b$ and consequently $\sigma_j(D) = \sigma_j(D')$ for all $j \geqslant 0$. ∎

The proof of Proposition 7.15. Suppose first that l is odd. Let L be an arbitrary extension field of K over which $x^l - 1$ splits into linear factors i.e., $(x^l - 1) = \prod_{i=1}^{l}(x - \epsilon_i)$ holds with $\epsilon_i \in L$. Note that the ϵ_i are not necessarily distinct but $\sigma_j(\epsilon_1,...,\epsilon_l) = 0$ holds for all $j > 0$ except for $j = l$, where one has $\sigma_l(\epsilon_1,...,\epsilon_l) = 1$. Let us compute $D_l(\epsilon_1,...,\epsilon_l,0,0,...,0)$. Denoting by d_{ij} the (i,j)-th entry of $D_l(\epsilon_1,...,\epsilon_l,0,...,0)$ we see that

$$d_{ij} = \begin{cases} 1 & \text{if } j \equiv 2i \ (\text{mod } l) \\ 0 & \text{otherwise} \end{cases}.$$

Consequently $\det D_l(\epsilon_1,...,\epsilon_l,0,...,0) = (-1)^{\left[\binom{(l+1)/2}{2}\right]} \neq 0$, thus $\det D_l(X)$ is not the zero-polynomial over K either.

If l is even, we argue in the same way, but for an extension field L', containing all the roots of $x^{l+1}-1$. ∎

The proof of Proposition 7.16. Fix a proper, non-empty subset $N = \{n_1, n_2, ..., n_j\} \subset \{1, 2, ..., k\}$. Let us suppose that for $F = \{x_1, ..., x_k\} \in \binom{X}{k}$, and $Y = \{x_i \in F, i \in N\}$, $\det D_l(Y) = 0$ ($|Y| = |N| > l$). By the preceding proposition we may find $n_i \in N$ for which $\det D_l(x_{n_1}, ..., x_{n_{i-1}}, y, x_{n_{i+1}}, ..., x_{n_j})$ considered as a polynomial, $p(y)$ of y is not identically zero. Moreover, $\deg p(y) \leqslant l$. For any choice of $\{x_1, x_2, ..., x_{n_{i-1}}, x_{n_i+1}, ..., x_j\}$ there are at most l values of y satisfying $\det D_l(Y) = p(y) = 0$. Since there are only $2^k - 2$ choices for N, and less than k choices for l, the statement follows. ∎

The proof of Proposition 7.17. Let us look at the values of $\sigma_{2i}(X)$ for $1 \leqslant i \leqslant t$, $|X| = k$ and X such that $\det D_l(Y) \neq 0$ holds for all $Y \subset X$. The number of possible value sequences is bounded by q^t. On the other hand, by Proposition 7.16 we have at least $\binom{q}{k} - 2^k k^2 \binom{q}{k-1}$ choices for X, thus there is a particular value sequence, say $c_2, c_4, ..., c_{2t}$ which occurs at least $\binom{q}{k}/q^t - 2^k k^2 \binom{q}{k-1}/q^t$ times. ∎

The proof of the lower bounds of (7.6) and (7.7). First we prove (7.6). Let $\mathscr{F} \subset \binom{X}{3t+1}$ be a family of maximal size and without four distinct sets A, B, A', B' satisfying $A \cup B = A' \cup B'$, $A \cap B = A' \cap B'$. Define $\mathscr{F}(x) = \{F - \{x\} : x \in F \in \mathscr{F}\}$, for every $x \in X$. Obviously $\sum_{x \in X} |\mathscr{F}(x)| = (3t+1)|\mathscr{F}| = (3t+1)v_{3t+1}(n)$ holds. Thus we may choose some $x \in X$ satisfying $|\mathscr{F}(x)| \geqslant \frac{3t+1}{n} v_{3t+1}(n)$. Moreover, $\mathscr{F}(x)$ fulfils the assumptions: there are no four distinct sets $A, A', B, B' \in \mathscr{F}(x)$ satisfying $A \cup B = A' \cup B'$, $A \cap B = A' \cap B'$. Moreover, $\mathscr{F}(x) \subset \binom{X-\{x\}}{3t}$.

We infer

$$\nu_{3t}(n) \geq |\mathcal{F}x)| \geq \frac{3t+1}{n}\left[\frac{1}{(3t+1)!} - o(1)\right]n^{2t+1} = \left[\frac{1}{(3t)!} - o(1)\right]n^{2t}.$$

To prove (7.7) consider again a family $\mathcal{F}$ realizing the maximum size but this time for $k' = 2k = 6t+4 = 3(2t+1)+1$. For each $F \in \mathcal{F}$ let us fix a partition of $\mathcal{F}$ into 2-element sets: $F = P_1 \cup \cdots \cup P_{3t+2}$. Let us define

$$\tilde{\mathcal{F}} = \{\{P_1, ..., P_{3t+2}\} : P_i \in \begin{bmatrix}X\\2\end{bmatrix}, \ 1 \leq i \leq 3t+2, \ (P_1 \cup ... \cup P_{3t+2}) \in \mathcal{F}\}.$$

Obviously $|\tilde{\mathcal{F}}| = |\mathcal{F}|$, and $\tilde{\mathcal{F}}$ satisfies that for any four distinct $A, B, A', B' \in \tilde{\mathcal{F}}$ either $A \cup B \neq A' \cup B'$ or $A \cap B \neq A' \cap B'$. Thus, we have $\nu_{3t+2}\left[\begin{bmatrix}n\\2\end{bmatrix}\right] \geq \frac{1-o(1)}{(6t+4)!}\left[2\begin{bmatrix}n\\2\end{bmatrix}\right]^{(4t+3)/2}$

Using (7.5) and setting $3t+2 = k$ we infer

$$\nu_k(n) \geq \left[\frac{1}{2k!} - o(1)\right](2n)^{\frac{4t+3}{2}}. \quad \blacksquare$$

8. Union-Free Families without Size Restrictions

Suppose $X = X_1 \oplus X_2 \oplus \cdots \oplus X_t$ and $\mathcal{F}$ is the complete t-partite t-graph with classes $X_1, ..., X_t$, i.e.,

$$\mathcal{F} = \{F \subset X : |F \cap X_i| = 1, \ i = 1, ..., t\}.$$

Then clearly $|\mathcal{F}| = \prod_{i=1}^{t} |X_i|$ and $\mathcal{F}$ is cancellative. It is easy to see that among such families $|\mathcal{F}|$ is maximal if and only if all X_i have size 3 with the exception of at most two which have size 2.

Thus, denoting by $c(n)$ the maximum-size of a cancellative family we see

$$c(n) \geq \begin{cases} 3^{n/3} & \text{if} \quad n \equiv 0 \pmod 3 \\ 3^{(n-4)/3} \cdot 2^2 & \text{if} \quad n \equiv 1 \pmod 3 \\ 3^{(n-2)/3} \cdot 2 & \text{if} \quad n \equiv 2 \pmod 3 \end{cases} \tag{8.1}$$

Erdös and Katona (*cf.* [K]) conjecture that in (8.1) equality holds.

Theorem 8.1. (Frankl and Füredi [FF3]

$$c(n) \leq \sqrt{2n} \cdot 1.5^n.$$

Proof of Theorem 8.1. Set $\mathscr{F}^k = \{F \in \mathscr{F} : |F| = k\}$ and $f_k = |\mathscr{F}^k|$. Note that Proposition 7.10 implies for $k \geqslant \dfrac{n}{2}$

$$f_k \leqslant 2^{n-k} \tag{8.2}$$

Next we bound f_k for $k < \dfrac{n}{2}$.

Suppose $Y \in \binom{X}{2k}$ and $\mathscr{G}(Y) = \{F \in \mathscr{F}^k : F \subset Y\}$. In view of Proposition 7.10 $|\mathscr{G}(Y)| \leqslant 2^k$.

Since $|\mathscr{F}^k| = \dfrac{1}{\binom{n-k}{k}} \displaystyle\sum_{Y \in \binom{X}{2k}} |\mathscr{G}(Y)|$, we infer

$$f_k \leqslant \binom{n}{k} 2^k \Big/ \binom{2k}{k}.$$

This and $\binom{2k}{k} \geqslant 2^{2k} / \sqrt{4k}$ (prove this by induction) imply

$$f_k \leqslant \sqrt{2n} \, \binom{n}{k} 2^{-k}. \tag{8.3}$$

Clearly the RHS of (8.2) is less than that of (8.3), thus we infer summing (8.3) over all k

$$|\mathscr{F}| \leqslant \sqrt{2n} \, \sum_k \binom{n}{k} 2^{-k} = \sqrt{2n} \, \left[\frac{3}{2}\right]^n. \quad \blacksquare$$

The problem of determining the maximum size of (strongly) union-free families, which we denote by $u(n)$ goes back to Erdös and Moser [Calgary Conf. 1969]. They observed that if $\mathscr{F} = \{F_1, ..., F_m\} \subset 2^X$ is strongly union-free then all the $\binom{m}{2}$ unions $F_i \cup F_j$, are distinct and thus $\binom{m}{2} \leqslant 2^n$ holds. This implies

$$u(n) \leqslant 2^{(n+1)/2} + 1. \tag{8.4}$$

They mention also that an exponential lower bound can be found by a random "construction".

Here we exhibit a slightly weaker but explicit construction.

Suppose $n \geqslant 2t$ and Q_1, Q_2 are two disjoint copies of the finite field of 2^t elements — we write the same symbols for the corresponding elements of Q_1 and Q_2. Note that the elements of Q_1, Q_2 are $(0,1)$-sequences of length t, i.e., they can be represented as subsets of $\{1, 2, ..., t\}$, $\{t+1, ..., 2t\}$, respectively. For $x_1 \in Q_1$, $x_2 \in Q_2$ denote by $F(x_1)$, $F(x_2)$ its set representation.

Theorem 8.2. (Lindström [Li2]). In the family $\mathscr{S} = \{F(x) \cup F(x^3) : x \in Q_1, x^3 \in Q_2\}$ all pairwise symmetric differences are distinct.

Proof. First we note that $F(x) \Delta F(y) = F(x+y)$. Thus we must show that $x+y = a+b$, $x^3 + y^3 = a^3 + b^3$ imply $\{x,y\} = \{a,b\}$ if $x \neq y$. The second equation can be written as

$$(x+y)((x+y)^2 - xy) = (a+b)((a+b)^2 - ab).$$

Since $x \neq y$, $x+y \neq 0$, using the first equation we infer $xy = ab$, that is x,y and a,b have their symmetric polynomials equal. Therefore $\{x,y\} = \{a,b\}$ (both are the set of roots of the polynomial $z^2 - (x+y)z + xy$ in Q_1). ∎

Now suppose $n \geqslant 4t$ and take four copies Q_1, Q_2, Q_3, Q_4 of the finite field of 2^t elements along with their set representations on $[1,t]$, $[t+1,2t]$, $[2t+1,3t]$, $[3t+1,4t]$, respectively. Denote by $\bar{1}$ the element $(1,1,...,1)$ of the field.

Proposition 8.3. ([FF3]). The family

$$\mathscr{F} = \{F_1(x) \cup F_2(x^3) \cup F_3(x+\bar{1}) \cup F_4(x^3+\bar{1})\}$$

is union-free.

Proof. Suppose x,y,a,b are elements of the field such that $F_1(x) \cup F_1(y) = F_1(a) \cup F_1(b)$ and $F_3(x+\bar{1}) \cup F_3(y+\bar{1}) = F_3(a+\bar{1}) \cup F_3(b+\bar{1})$.

Since $F_3(z+\bar{1})$ is the complement (in $\{2t+1,...,3t\}$) of $F_3(z)$, the second equality implies that $F_1(x) \cap F_1(y) = F_1(a) \cap F_1(b)$. Consequently $F_1(x) \Delta F_1(y) = F_1(a) \Delta F_1(b)$, and similarly for the third powers. Thus Theorem 8.2 implies $\{x,y\} = \{a,b\}$. ∎

(8.4) and Proposition 8.3 yield

Theorem 8.4.

$$2^{n/4(1+o(1))} \leqslant u(n) \leqslant \sqrt{2}\, 2^{n/2} + 1.$$

Finally, let us consider the maximum size, say $w(n)$ $(v(n))$ of a weakly union-free (union- or intersection-free) family $\mathscr{F} \subset 2^X$, respectively. The best upper bound is due to Lindström [Li1].

Theorem 8.5. (Lindström) For $n > n_0$ one has

$$w(n) \leqslant v(n) < 3 \cdot 2^{3n/5}.$$

Proof. Let us represent the members of $\mathscr{F}$ by $(0,1)$-vectors of length n, say $v(F)$ is the vector corresponding to F.

Proposition 8.6. If $\mathscr{F}$ is union- or intersection-free than all the vectors $v(F) - v(F')$ are distinct $(-1,0,1)$-vectors, $F, F' \in \mathscr{F}$, F and F' are distinct.

Proof. Consider $v(F) + v(G)$ then clearly $i \in F \cap G$ if and only if the i'th coordinate is 2, $i \in F \cup F'$ if and only if the i'th coordinate is nonzero. Thus $v(F) + v(G)$ determines $F \cap G$ and $F \cup G$. Since $v(F) - v(F') = v(G') - v(G)$ implies $v(F) + v(G) = v(F') + v(G')$, the statement follows. $\blacksquare$

We need a simple observation.

Proposition 8.7. Let $u_1, \ldots, u_m$ be $(0,1)$-vectors of length n and let t be a fixed integer, $1 \leqslant t \leqslant n$. Then at least half of the m^2 differences $u_i - u_j$, $1 \leqslant i \neq j \leqslant m$, have a 0 in the t'th position.

Proof. Let us suppose that there are a vectors among the u_i which have 0 in the t'th position. Then out of the differences $a^2 + (m-a)^2 \geqslant \dfrac{1}{2} m^2$ will have a 0 there. $\blacksquare$

Let us fix two positive integers e, f with $e + f = n$. Fix a partition $X = Y \oplus Z$ with $|Y| = e$, $|Z| = f$.

Let $\mathscr{F} \subset 2^X$ be intersection or union-free.

For $F \in \mathscr{F}$ let $w(F)$ be the characteristic vector of $F \cap Z$.

For $A \subset Y$ define

$$W(A) = \{w(F): F \in \mathscr{F}, F \cap Y = A\}.$$

Note that we have

$$\sum_{A \subset Y} |W(A)| = |\mathscr{A}| \ .$$

There are 2^e terms on the LHS. Thus the inequality between arithmetic and quadratic means gives

$$\sum_{A \subset Y} |W(A)|^2 \geqslant |\mathscr{A}|^2 / 2^e \ . \tag{8.6}$$

Note that the LHS of (8.6) is the total number of vectors $w(F) - w(F')$ where $F, F' \in \mathscr{F}$, $F \cap Y = F' \cap Y$. Thus the zero vector is counted $|\mathscr{A}|$-times. By Proposition 8.6 all other vectors occur at most once.

Let us suppose for simplicity f is even, say $f = 2g$.

Let h_r denote the number of vectors among the differences which have exactly r zeros. Then we have by (8.6) and by Proposition 8.7

$$\sum_r h_r \geqslant |\mathscr{A}|^2 / 2^e \tag{8.7}$$

$$\sum_r r \, h_r \geqslant g \sum_r h_r \ , \text{ and consequently}$$

$$\sum_{r < g} h_r \leqslant \sum_{r > g} (r-g) h_r \tag{8.8}$$

Since $h_{2g} = |\mathscr{A}|$ and $h_r \leqslant \binom{f}{r} 2^{f-r}$ for $r < 2g$, (8.8) implies

$$\sum_{r < g} h_r \leqslant g |\mathscr{A}| + \sum_{g < r < 2g} (r-g) \binom{2g}{r} 2^{2g-r} \tag{8.9}$$

Using (8.7) and (8.9) we obtain

$$2^{-e} |\mathscr{A}|^2 - (g+1) |\mathscr{A}| \leqslant 2^g \sum_{0 \leqslant i \leqslant g} (i+1) \binom{2g}{g+i} 2^{-i} \leqslant 2^g \binom{2g}{g} \sum_{0 \leqslant i \leqslant g} \frac{i+1}{2^i} = 2^g \binom{2g}{g} \cdot 4 \ .$$

Thus we have proved, that for all positive integers e, g with $e + 2g = n$ one has

$$|\mathscr{A}|^2 \leqslant 2^e (g+1) |\mathscr{A}| + 2^{e+g+2} \binom{2g}{g} \ .$$

Choosing $g \sim \left\lfloor \dfrac{n}{5} \right\rfloor$ to optimize, gives for n sufficiently large $|\mathscr{A}| < 3 \cdot 2^{3n/5}$, as desired. $\blacksquare$

By probabilistic methods one can prove exponential lower bounds for both $w(n)$ and $v(n)$.

As an example we prove

$$v(n) > \frac{1}{2}(8/3)^{n/3}. \tag{8.10}$$

Proof of (8.10). Choose, at random, subsets of $\{1, 2, ..., n\}$, that is the sets are chosen independently and all 2^n subsets are equiprobable. The event that the element i does not spoil the condition $F_1 \cup F_2 = F_3 \cup F_4$ and $F_1 \cap F_2 = F_3 \cap F_4$ is equivalent to $i \in F_1 \cap F_2 \cap F_3 \cap F_4$ or $i \notin F_1 \cup F_2 \cup F_3 \cup F_4$ or $i \in ((F_1 \Delta F_2) \cap (F_3 \Delta F_4))$. It has probability $\frac{1}{16} + \frac{1}{16} + \frac{1}{4} = \frac{3}{8}$. Thus the probability that four randomly chosen sets don't form an intersection- or union-free family is $3 \cdot (3/8)^n$ (the factor 3 comes from the 3 ways of pairing the sets). Thus the expected number of 4-tuples violating the desired property is $\binom{m}{4} 3 \cdot (3/8)^n$. Removing one set out of each bad 4-tuple leaves $m - \binom{m}{4} 3 \cdot (3/8)^n$ sets, and then there can be at most one set which appears twice. Choosing $m = \sqrt[3]{(8/3)^n/2}$, the expected number of bad 4-tuples is less than $\frac{m}{4}$, thus we obtain an intersection- or union-free family of size at least $3m/4 - 1$. ∎

For much weaker but still exponential explicit constructions see Alon [A].

9. No Large Boolean Subalgebra

A Boolean algebra of dimension d is a family $\mathcal{B}$ of 2^d sets which behave for intersection and union as the subsets of a d-element set. That is, there exist $d+1$ pairwise disjoint sets $A_0, A_1, ..., A_d$ so that

$$\mathcal{B} = \{A_0 \cup A_{i_1} \cup \cdots \cup A_{i_s} : 1 \leqslant i_1 < \cdots < i_s \leqslant d\}.$$

So a Boolean algebra of dimension 1 is just a pair of two sets one containing the other.

Let us define $b(n, d)$ as the maximum size of a family of subsets of an n-set without a Boolean algebra of dimension d.

Thus the classical Sperner Theorem is equivalent to

$$b(n,1) = \left(\!\left[\begin{array}{c} n \\ \left\lfloor\frac{n}{2}\right\rfloor \end{array}\right]\!\right) = O(2^n/\sqrt{n}).$$

Theorem 9.1. (Rödl [R])

$$b(n,d) = O(2^n/n^{2^{-d}}).$$

Remark 9.2. The case $d=2$ was already solved by Erdös and Kleitman [EK1] who showed that actually $b(n,2) = \Theta(2^n/n^{1/4})$ holds by providing an appropriate lower bound. In general it is not decided whether Rödl's bound is sharp. Let us mention that $b(n,d) = o(2^n)$ was also shown by Brown and Buhler [BB].

Before giving the proof of Theorem 9.1 let us recall that given a set $Y, |Y| = t$, 2^Y has a symmetric chain decomposition (cf [K]) into $\left(\!\left[\begin{array}{c} t \\ \left\lfloor\frac{t}{2}\right\rfloor \end{array}\right]\!\right)$ chains where the number of chains of length at most $2i$ is $\left(\!\left[\begin{array}{c} t \\ \left\lfloor\frac{t}{2}\right\rfloor \end{array}\right]\!\right) - \left(\!\left[\begin{array}{c} t \\ \left\lfloor\frac{t}{2}\right\rfloor - i \end{array}\right]\!\right)$ (because the remaining chains contain one $\left(\!\left[\left\lfloor\frac{t}{2}\right\rfloor - i\right]\!\right)$-subset, each). Suppose for definiteness $Y = \{y_1,...,y_t\}$ and $\mathscr{S}_1,...,\mathscr{S}_{\left(\!\left[\begin{array}{c} t \\ \left\lfloor\frac{t}{2}\right\rfloor - i \end{array}\right]\!\right)}$ are the disjoint chains containing $\left(\!\left[\left\lfloor\frac{t}{2}\right\rfloor - i\right]\!\right)$-element sets. If π is a permutation of $\{1,2,...,t\}$ then $\pi(\mathscr{S}_1), \pi(\mathscr{S}_2), \cdots$ is another system of chains where if $C_j \in \mathscr{S}_j$ and $C_j = \{y_{i_1},...,y_{i_r}\}$ then $\pi(C_j) = \{y_{\pi(i_1)},...,y_{\pi(i_r)}\}$ and $\pi(\mathscr{S}_j) = \{\pi(C_i) : C_i \in \mathscr{S}_i\}$. Let us set $\mathscr{S} = \bigcup\limits_{1 \leqslant j \leqslant \left(\!\left[\begin{array}{c} t \\ \left\lfloor\frac{t}{2}\right\rfloor - i \end{array}\right]\!\right)} \mathscr{S}_j$.

Proposition 9.4. If π is a random permutation of $\{1,2,...,t\}$ and D is an arbitrary subset of Y then the probability of $D \in \pi(\mathscr{S})$ satisfies

$$p(D \in \pi(\mathscr{S})) \geqslant \left(\!\left[\begin{array}{c} t \\ \left\lfloor\frac{t}{2}\right\rfloor - i \end{array}\right]\!\right) \Big/ \left(\!\left[\begin{array}{c} t \\ \left\lfloor\frac{t}{2}\right\rfloor \end{array}\right]\!\right). \tag{9.1}$$

Proof. If $|D| \leqslant \left\lfloor\frac{t}{2}\right\rfloor - i$ or $|D| \geqslant \left\lfloor\frac{t}{2}\right\rfloor + i$ then clearly the probability in question is 1 because $\mathscr{S}$

and thus $\pi(\mathscr{S})$ contain all sets of these sizes. If $\left\lfloor\frac{t}{2}\right\rfloor - i \leqslant |D| \leqslant \left\lceil\frac{t}{2}\right\rceil + i$, then clearly

$$p(D \in \pi(\mathscr{S})) = \binom{\left\lfloor\frac{t}{2}\right\rfloor - i}{t} \bigg/ \binom{t}{|D|} \geqslant \binom{\left\lfloor\frac{t}{2}\right\rfloor - i}{t} \bigg/ \binom{t}{\left\lfloor\frac{t}{2}\right\rfloor}. \quad \blacksquare$$

Let us estimate the RHS of (9.1)

$$\binom{\left\lfloor\frac{t}{2}\right\rfloor - i}{t} \bigg/ \binom{\left\lfloor\frac{t}{2}\right\rfloor}{t} = \frac{\left\lfloor\frac{t}{2}\right\rfloor - i + 1}{\left\lfloor\frac{t}{2}\right\rfloor + 1} \cdots \frac{\left\lfloor\frac{t}{2}\right\rfloor}{\left\lceil\frac{t}{2}\right\rceil + i} > \left(1 - \frac{2i+2}{t}\right)^i. \tag{9.2}$$

Proof of Theorem 9.1. Let now $X = X_1 \oplus X_2 \oplus \cdots \oplus X_d$ be an arbitrary partition of X, satisfying $\left\lfloor\frac{n}{d}\right\rfloor \leqslant |X_j| \leqslant \left\lceil\frac{n}{d}\right\rceil$. Let $\mathscr{S}$ be the union of the chains of length at least $2i$ in a symmetric chain decomposition of X_j (i will be fixed later), and let π_j be a random permutation of $\{1,2,...,|X_j|\}$.

Suppose $\mathscr{F} \subset 2^X$, $\mathscr{F}$ contains no d-dimensional Boolean algebra. For an arbitrary $F \in \mathscr{F}$ Proposition 9.4 and (9.2) imply $p(F \cap X_j \in \pi_j(\mathscr{S})$ holds for $j = 1,...,d) > \left(1 - \frac{2i+2}{\left\lfloor\frac{n}{d}\right\rfloor}\right)^{id}$. E.g., for

$i = \left\lfloor\sqrt{n/d}\right\rfloor$ and $n > n_0(d)$ this probability is greater than 0.1^d. Thus the expected number of $F \in \mathscr{F}$ with $F \cap X_j \in \pi_j(\mathscr{S})$, $j = 1,...,d$ is greater than $0.1^d|\mathscr{F}|$. Therefore there must be a choice of $\pi_1,...,\pi_d$ for which the number of such $F \in \mathscr{F}$ is at least as much. Let $\mathscr{D}^1,...,\mathscr{D}^d$ be the corresponding families of symmetric chains.

If $\mathscr{D}^j_{l_j}$ is a chain in $\mathscr{D}^j$, $j = 1,...,d$ then define

$$\mathscr{D}^1_{l_1} \otimes \cdots \otimes \mathscr{D}^d_{l_d} = \left\{\bigoplus_{j=1}^{d} D^j_{l_j} : D^j_{l_j} \in \mathscr{D}^j_{l_j}\right\}.$$

Set also

$$\mathscr{D}^1 \otimes \cdots \otimes \mathscr{D}^d = \bigoplus_{l_1,l_2,...,l_d} (\mathscr{D}^1_{l_1} \otimes \cdots \otimes D^d_{l_d}).$$

Then clearly for some choice of $l_1,l_2,...,l_d$

$$\frac{\left|\mathcal{F}\cap\left(\mathcal{D}_{l_1}^1\otimes\cdots\otimes\mathcal{D}_{l_d}^d\right)\right|}{\left|\mathcal{D}_{l_1}^1\right|\cdots\left|\mathcal{D}_{l_d}^d\right|} \geqslant \frac{\left|\mathcal{F}\cap\left(\mathcal{D}^1\otimes\cdots\otimes\mathcal{D}^d\right)\right|}{\left|\mathcal{D}^1\right|\cdots\left|\mathcal{D}^d\right|} > \frac{0.1^d|\mathcal{F}|}{2^n}.$$

Thus Theorem 9.1 will follow if we show that the LHS is bounded from above by $9/(2i+1)^{2^{1-d}} < c/n^{2^{-d}}$ with some constant $c = c(d)$. However, the contrary would imply in view of a theorem of Erdös, [E2] that for some $l_1,...,l_d$ there exist $D_{l_j}^j, E_{l_j}^j \in \mathcal{D}_l^j$ such that all the 2^d sets F satisfying $F \cap X_j \in \{D_{l_j}^j, E_{l_j}^j\}$, $j = 1,...,d$ are in $\mathcal{F}$. Since $\mathcal{D}_l^j$ is a chain we may suppose $D_{l_j}^j \subsetneqq E_{l_j}^j$ and set $A_j = E_{l_j}^j - D_{l_j}^j$, $A_0 = D_{l_1}^1 \cup \cdots \cup D_{l_d}^d$. That is, the d-dimensional Boolean algebra $\{A_0 \cup A_{i_1} \cup \cdots \cup A_{i_s} : 1 \leqslant i_1 < \cdots < i_s \leqslant d\}$ is contained in $\mathcal{F}$, a contradiction. ∎

References

[A] N. Alon, Explicit construction of exponential sized families of k-independent sets, Discrete Math. 58 (1986) 191-193.

[AF] N. Alon and P. Frankl, Families in which disjoint sets have large unions, Annals New York Acad. Sci., to appear.

[B] B. Bollobás, Three-graphs without two triples whose symmetric difference is contained in a third, Discrete Math. 8 (1974), 21-24.

[BB] T. C. Brown and J. P. Buhler, Lines imply spaces in density Ramsey theory, J. Combinatorial Th. A 36 (1984), 214-220.

[DL] D. E. Daykin and L. Lovász, The number of values of a Boolean function, J. London Math. Soc. 12 (1975), 225-230.

[E1] P. Erdös, On sequences of integers no one of which divides the product of two others, Mitt. Forsch. Inst. Math. Mech., Tomsk 2 (1938), 74-82.

[E2] P. Erdös, On extremal problems of graphs and generalized graphs, Israel J. Math. 2 (1964), 183-190.

[EFF1] P. Erdös, P. Frankl and Z. Füredi, On families of sets in which no set is covered by the union of two others, J. Combinatorial Th. A 33 (1982), 158-166.

[EFF2] P. Erdös, P. Frankl and Z. Füredi, Families of finite sets in which no set is covered by the union of r others, Israel J. Math. 51 (1985), 79-89.

[EK1] P. Erdös and D. J. Kleitman, On collections of subsets containing no 4-member Boolean algebra, Proc. AMS 28 (1971), 87-90.

[EK2] P. Erdös and D. J. Kleitman, On colorings of graphs to maximize the proportion of multicolored k-edges, J. Combinatorial Th. 5 (1968), 164-169.

[EKR] P. Erdös, C. Ko and R. Rado, Intersection theorems for systems of finite sets, Quart. J. Math. Oxford 12 (1961), 313-320.

[ER] P. Erdös and A. Rényi, On a problem in the theory of graphs, Publ. Math. Inst. Hung. Acad. Sci. 7A (1962), 623-641, in Hungarian.

[F1] P. Frankl, Families of finite sets containing no k disjoint sets and their union, Periodica Math. Hung. 8 (1977), 29-31.

[F2] P. Frankl, Intersection theorems for finite sets and geometric applications, Proc. Internat. Congress of Math., Berkeley, 1986.

[FF1] P. Frankl and Z. Füredi, A new extremal property of Steiner triple systems, Discrete Math. 48 (1984), 205-212.

[FF2] P. Frankl and Z. Füredi, Union free families of sets and equations over fields, J. Number Th. 23 (1986), 210-218.

[FF3] P. Frankl and Z. Füredi, Union free families and probability theory, European J. Comb. 5 (1984), 127-131, Errata, ibid. p. 395.

[Fü1] Z. Füredi, Graphs without quadrilaterals, J. Combinatorial Th. B 34 (1983), 187-190.

[Fü2] Z. Füredi, Hypergraphs in which all disjoint pairs have distinct unions, Combinatorica 4 (1984), 161-168.

[Fü3] Z. Füredi, Extremal problems for the principal parameters of hypergraphs, Graphs and Comb. 4 (1988).

[HM] A. J. W. Hilton and E. C. Milner, Some intersection theorems for systems of finite sets, Quart. J. Math. Oxford 18 (1967), 369-384.

[K] G. O. H. Katona, Extremal problems for hypergraphs, in Combinatorics Vol II. pp. 13-42 (M. Hall, J. H. van Lint, Eds.) Math. Centre Tracts 56 (1974), Amsterdam.

[KSS] D. J. Kleitman, J. B. Shearer and D. Sturtevant, Intersections of k-element sets,

Combinatorica 1 (1981), 381-384.

[KST] T. Kövári, V. T. Sós and P. Turán, On a problem of K. Zarankiewicz, Colloq. Math. 3 (1959), 50-57.

[KT] G. O. H. Katona and T. G. Tarjan, Extremal problems with excluded subgraphs in the n-cube. Graph theory (Lagów, 1981), 84-93, Lecture Notes in Math., 1018, Springer, Berlin-New York, 1983.

[Kl1] D. J. Kleitman, Families of non-disjoint subsets, J. Combinatorial Th. 1 (1966), 153-155.

[Kl2] D. J. Kleitman, Maximal number of subsets of a finite set no k of which are pairwise disjoint, J. Combinatorial Th. 5 (1968), 157-163.

[Kl3] D. J. Kleitman, On families of subsets of a finite set containing no two disjoint sets and their union, J. Combinatorial Th. 5 (1968), 235-237.

[Kl4] D. J. Kleitman, Extremal properties of collections of subsets containing no two sets and their union, J. Combinatorial Th. A 20 (1976), 390-392.

[Kl5] D. J. Kleitman, Collections of sets without unions and intersections, Annals New York Acad. Sci., to appear.

[Li1] B. Lindström, On B_2-sequences of vectors, J. Number Th. 4 (1972), 261-265.

[Li2] B. Lindström, Determination of two vectors from the sum, J. Combinatorial Th. 6 (1969), 402-407.

[L] L. Lovász, Combinatorial problems and exercises, North Holland, Amsterdam, 1979.

[M] W. Mantel, Problem 28, Wiskundige Opgaven 10 (1907), 60-61.

[Mc] R. J. McEliece, The theory of information and coding, Addison-Wesley, Cambridge, 1977.

[MS] J. Marica and J. Schönheim, Differences of sets and a problem of Graham, Canad. Math. Bull. 12 (1969), 635-637.

[R1] V. Rödl, On a packing and covering problem, European J. Comb. 6 (1985), 69-78.

[R2] V. Rödl, Families of sets without large Boolean algebras, manuscript.

[Si] F. Sidorenko, On a problem of Bollobas on 4-graphs, Mat. Zametki, to appear, in Russian.

[S] E. Sperner, Ein Satz über Untermengen einer endlichen Menge, Math. Z. 27 (1928), 544-548.

DISTANCE-TRANSITIVE GRAPHS OF VALENCY k, $8 \leqslant k \leqslant 13$

A.A. Ivanov
Institute for System Studies, Academy of Sciences
of the USSR, 9, Prospect 60 Let Oktyabrya,
117312, Moscow, USSR

A.V. Ivanov
Institute for System Studies, Academy of Sciences
of the USSR, 9, Prospect 60 Let Oktyabrya,
117312, Moscow, USSR

1 INTRODUCTION

In Faradžev et al. (1986) an algorithm for
enumeration of intersection arrays of distance-transitive
graphs (d.t.g.) having given valency k is described. This
algorithm was used in Faradžev et al. (1986) for classification of the d.t.g.'s of valency k = 5, 6 and 7. This
paper is a logical continuation of Faradžev et al. (1986).
Here we supply the algorithm with some new feasibility conditions. The new version of the algorithm enable us to
classify the d.t.g.'s of valency up to 13.

All definitions and notations concerning
d.t.g.'s can be found in Faradžev et al. (1986) (see also
Bannai & Ito (1984) and Brouwer et al. (1987)).

Let us summarize briefly the history of classification of d.t.g.'s having small valences. Chronological-
ly the first result in the area is the classification of
cubic d.t.g.'s (k=3), obtained by Biggs & Smith (1971).
Their scheme of classification can be generalized on the
case of an arbitrary valency $k \geqslant 3$ and in general consists
of the following four steps:

(1) to prove that the order of the vertex
stabilizer in the automorphism group of d.t.g. having valency k is bounded by some value f = f(k);

(2) to bound the diameter of a d.t.g. having
valency k by some value D = D(k);

(3) to enumerate all feasible intersection arrays for the given valency k;

(4) to construct all graphs with intersection
arrays found on the step (3).

It is easy to see that if the step (2) is done, the number of arrays one has to consider in order to realize the step (3) is bounded by certain function of k.

For k = 3 the step (1) was done by Tutte (1947) and Sims (1967), on the step (2) the realization of the previous one was used essentially and on the step (3) a computer calculations were involved.

For k = 4 the classification was carried out by Smith (1973, 1974 a,b). On the first step he used the results of Sims (1968), Quirin (1971) and Gardiner (1973).

A few years ago Cameron (1982) and A.A.Ivanov (1983) have proved independently that the step (2) follows from the first one for all $k \geqslant 3$. In other words it was proved that the diameter of a d.t.g. having valency $k \geqslant 3$ is bounded by certain function of the order of the vertex stabilizer in its automorphism group.

This bounding of the diameter and some recent results (Weiss(1981 a,b);Goldschmidt(1980)) concerning edge-transitive graphs enable Faradžev et al. (1986) to classify d.t.g.'s having valency 5, 6 and 7. Their scheme of classification was very similar to the above one.

Recently the effectivity of this approach was proved for all $k \geqslant 3$. Namely, the step (1) was realized at first by Cameron et al. (1983) modulo the classification of the finite simple groups and then by Weiss (1985) (see also section 6.3 in Brouwer et al. (1987)) without any assumptions.In Cameron et al. (1983) the function f(k) was described unconstructively. From Weiss (1985) this function can be calculated directly, but it grows up very rapidly with increasing of k. So these results give the possibility to classify d.t.g.'s of an arbitrary valency k in principle but the realization of the step (3) demands a great number of calculations. This makes it necessary to look for an alternative approach.

At the same time another scheme of classification of d.t.g.'s was developing. This scheme was proposed by Gardiner (1975), where an independent classification of d.t.g.'s of valency 3 was obtained. He has realized the

step (1) on a more detailed level. Namely, all possible
vertex stabilizers in automorphism groups of d.t.g.'s having
valency 3 were described up to isomorphism .This enable A.
Gardiner to do the steps (2) - (4) simultaneously and without computer calculations, but using essentially the information about vertex stabilizers. The description of the primitive permutation groups having subdegree 3, obtained by
Wong (1967), have played a significant role in the classification. In Gardiner (1982) an analogous strategy in the
case of arbitrary valency k was proposed. This strategy was
realized for k = 4 by Gardiner (1985). By Gardiner & Praeger (1987, 1986) the d.t.g.'s of valency 5 and 6 have been
classified in this way. A.A.Ivanov & Chuvaeva (1985) have
classified the d.t.g.'s of valency 11 using a modification
of the Gardiner's scheme. They used the fixed points method,
consisting in the following. During the construction of a
d.t.g. Γ with a given vertex stabilizer G(x) of vertex x
in the automorphism group G = Aut(Γ) a subgroup F $\leqslant$ G(x) is
considered. For the subgroup F the set of vertices fixed by
F is determined and the subgraph of Γ induced by these
vertices is reconstructed. If the subgroup F possesses some
special properties, the subgraph induced by the vertices
fixed by F is turned to be distance-transitive. This fact
enable to classify d.t.g. 's inductively. For example in
A.A.Ivanov & Chuvaeva (1985) during the classification of
d.t.g.' of valency 11 on the base of the fixed points method a detailed information about cubic d.t.g.'s was
used.

It should be noted that the approach of A.Gar-
diner in general and the fixed points method in particular
are effective in the cases when the order of G(x) is
"large" compared with k. On the other hand if the bound on
the order of G(x) is reduced, the computer enumeration of
intersection arrays can be carried out for sufficiently
large values of k. So we can use a complex approach to the
classification of d.t.g.'s. Namely, at first to characterize
without a computer the d.t.g.'s having "large" vertex stabilizer in the automorphism group and then after reduction of

the bound on the order of $G(x)$ to carry out a computer searching of the remaining intersection arrays. Just this
strategy is used in the paper.

As it was mentioned above, after some recent
results, the classification of d.t.g.'s of given valency
k becomes a purely technical problem. We have choosen 13
as the upper bound of considered valencies for the following reason.

In Biggs (1976) the results of computer investigation of automorphic graphs (d.t.g. with primitive
automorphism group) of valency $k \leqslant 13$ and diameter $d \leqslant 5$ are
presented. The classification of these graphs was reduced
in Biggs (1976) to the existence problem of graphs with
eleven given intersection arrays. The complete solution of
this problem was obtained in A.A.Ivanov (1981), Zaytchenko (1981), Zaytchenko et al. (1980, 1982) (see also the
survey Kalužnin et al. (1982)) and, independently in
Buekenhout & Rowlinson (1981), Cohen (1981), Gordon &
Levingstone (1981). So in this paper we repeat the classical result of N.Biggs but without any assumption about
diameter or primitivity. It is interesting to mention that
all automorphic graphs of valency $k \leqslant 13$ have appeared in
Biggs (1976) either in the form of known examples or in the
form of eleven open cases with given intersection arrays.

Let us give the content of the paper. Section
2 contains the characterizations of some significant classes of d.t.g.'s. In section 3 using these characterizations
upper bounds for the order of vertex stabilizer in the automorphism group of a d.t.g. of valency $8 \leqslant k \leqslant 13$ are obtained.
In section 4 some new combinatorial conditions included in
the algorithm are discussed. Section 5 contains some general results about imprimitive d.t.g.'s. Finally the section
6 contains the analysis of d.t.g.'s of valency $8 \leqslant k \leqslant 13$.

The authors are deeply indebted to I.A.Faradžev,
who has worked out the first version of computer program
for enumeration of intersection arrays and took part in
numerous discussion about it's modification.

We are highly acknowledged to A.M.Cohen for

pointing on the nonexistence of graph with intersection array N° 9 in Table 4.

2 CHARACTERIZATIONS

In this section we collect the characterizations of three classes of d.t.g.'s with "large" vertex stabilizer in the automorphism group.

The first class consists of the graphs with the property that the subgroup $G(x)$ induces on the set $\Gamma_1(x)$ the symmetric or the alternating group of the set. The characterization of this class of d.t.g. mainly follows from Cameron (1974), Praeger (1980) and Armanios (1981). The final form was firstly formulated in A.A.Ivanov (1984, 1986) and, independently, in Yokoyama (1986).

Theorem 2.1. Let Γ be a d.t.g. of valency $k \geqslant 6$ and the stabilizer $G(x)$ of a vertex $x \in V(\Gamma)$ in the automorphism group $G = \mathrm{Aut}(\Gamma)$ induce on $\Gamma_1(x)$ the symmetric or the alternating group of this set. Then either

i) Γ is isomorphic to a member of the following infinite series: K_k - the complete graphs; $K_{k,k}$ - the complete bipartite graphs; Q_k - the k-dimensional cubes; $\square_k$ - the folded graphs of the k-dimensional cubes; O_k - the odd graphs; $2.O_k$ - the bipartite double coverings of the odd graphs or

ii) $k = 7$ and Γ is the Hoffman - Singleton graph or the bipartite double covering of the latter graph.

The next class is the class of distance-transitive line graphs. Biggs (1974 b) has obtained the following result about these graphs.

Theorem 2.2. Let Γ be a d.t.g. of diameter $d \geqslant 2$ which is the line graph of a graph $\triangle$. Then $\triangle$ is a d.t.g. too, moreover $\triangle$ is either a Moore graph or the incidence graph of a generalized n-gon.

It is well known (see, for example, Biggs

(1974 a)),that a distance-transitive Moore graph is either
the Petersen graph or the Hoffman - Singleton graph. The
generalized n-gons whose incidence graphs are distance-transitive were classified in Gleason (1956), Higman (1964) and
Yanushka (1976).

Theorem 2.3. The generalized n-gons whose incidence graphs
are d.t.g. are classical generalized n-gons related to the
groups $A_2(p^m)$ (p is a prime), $B_2(2^m)$ and $G_2(3^m)$, $m \geqslant 1$.

Theorems 2.2 and 2.3 give the complete description of d.t.g.'s which are line graphs.

Very significant step in the classification of
d.t.g.'s was done by Weiss (1985 b). He has described all
d.t.g.'s which are s-transitive for $s \geqslant 4$.

Theorem 2.4. Let Γ be a d.t.g. of valency $k \geqslant 3$ and the
group Aut(Γ) act transitively on the set of paths of
length 4 in the graph Γ , then one of the following holds:

i) Γ is the incidence graph of a generalized
n-gon;

ii) $k = 3$, Γ has 102 vertices and Aut(Γ) $\cong$
$PSL_2(17)$;

iii) $k = 3$, Γ is the antipodal triple covering of the generalized 4-gon related to the group $B_2(2)$.

It should be mentioned that this theorem was
proved in Weiss (1985 b) under the assumption of the completeness of the known list of doubly transitive permutation
groups. Since the completeness of the list for small degrees k is proved by Sims (1970), our results do not depend
on this assumption. Weiss (1985 b) also contains the following significant consequence of theorem 2.4.

Theorem 2.5. Let Γ be a d.t.g. of valency $k \geqslant 3$ and girth
$g \geqslant 9$ then Γ is either the incidence graph of the generalized 6-gon related to the group $G_2(3^m)$ or one of the
graphs described in theorem 2.4 ii), iii).

In what follows we shall consider only the d.t.g.'s which distinct from the graphs characterized in this section.

3 BOUNDING THE ORDER OF VERTEX STABILIZER

For the enumeration of intersection arrays of d.t.g.'s it is necessary to find an upper bound for the value r, satisfying the following relations:

$$(1, a_1, b_1) = (c_{r-1}, a_{r-1}, b_{r-1}) \neq (c_r, a_r, b_r).$$

A general bound for this value r in terms of the valency k was obtained by Weiss (1985 b). But for concrete values of k this bound should be improved. A bound for r can be obtained by means of bounding the order of stabilizer $G(x)$ of a vertex x in the automorphism group $G = \mathrm{Aut}(\Gamma)$ of the graph Γ in question. In fact, since G acts distance-transitively on the graph Γ , the subgroup $G(x)$ acts transitively on the set $\Gamma_i(x)$ having cardinality k_i for all $0 \leq i \leq d$. In particular k_i divides $\left|G(x)\right|$. On the other hand $k_{r-1} = \left|\Gamma_{r-1}(x)\right| = k \cdot b_1^{r-2}$, hence $k \cdot b_1^{r-2}$ divides $G(x)$. So, having a bound for $\left|G(x)\right|$ we have a bound for r . Notice that if b_1 is not a power of a prime p, then to bound r it is sufficient to show that $\left|G(x)\right| = p^a \cdot q$ for a given integer q, such that $(p,q) = 1$. In this case we say that $\left|G(x)\right|$ divides $p^{\infty} \cdot q$.

We shall assume that $c_2 = 1$. This equality is always valid for $r \geqslant 3$. Moreover, if $c_2 \neq 1$ then some results of P.Terwilliger (see lemmas 4.14, 4.15 in section 4) enable to obtain quite strong bound for the diameter of a graph. So the intersection arrays can be effectively enumerated in this case.

Thus let $c_2 = 1$. Then each clique (a maximal complete subgraph) of Γ contains $(a_1 + 2)$ vertices and any two cliques have at most one vertex in common. This implies in particular that $(a_1 + 1)$ divides k . Let us consider the clique graph Δ of the graph Γ . By definition $V(\Delta)$ is the set of cliques of Γ , two vertices

from $V(\triangle)$ are adjacent in $\triangle$ if the cliques have a
nonempty intersection. It is easy to prove that Γ is the
clique graph of $\triangle$. If $k = 2\cdot(a_1 + 1)$ then every vertex of
Γ is contained in exactly two cliques and hence Γ is the
line graph of $\triangle$. Theorems 2.2 and 2.3 contain the complete description of the distance-transitive line graphs. So
we can assume that $k > 2\cdot(a_1 + 1)$.

For the values of k and a_1 in question
$(8 \leq k \leq 13$ and $(a_1 + 1)$ divides k) the information about
the order of $G(x)$ is presented in the Table 1. The table is
organized in the following way. If Γ is a d.t.g. of valen-
cy k, such that $c_2 = 1$, then $|G(x)|$ divides at least one
value in the corresponding item of the table. The meaning
of the values containing the symbol ∞ is defined above.

TABLE 1

$\diagdown\ a_1$ k	0	1	2	3
8	$2^9.3^2.7^2$	$2^\infty.3$; $2^6.3^\infty$		
9	$2^6.3^4.7^2$; $2^8.3^4$		$2^\infty.3^2$; $2^4.3^\infty$	
10	$2^9.3^6.5$	$2^{10}.3^4.5$; $2^{12}.5$		
11	$2^8.3^4.5^2.11$			
12	$2^{10}.3^5.5^2.11^2$	$2^{11}.3^4.5^2$; $2^\infty.3^2.5$; $2^8.3^\infty.5$	$2^7.3^4$; $2^4.3^6$	$2^{10}.3^5$; $2^{12}.3$
13	$2^8.3^\infty.13$			

Let us consider briefly the motivation of the table. If $a_1 = 0$ then for $x \in V(\Gamma)$ the subgroup $G(x)$ induces on $\Gamma_1(x)$ a doubly transitive permutation group $(G(x), \Gamma_1(x))$ and the action of G on Γ is s-transitive for $s \geqslant 2$. In view of theorem 2.1 we can assume that $(G(x), \Gamma_1(x))$ does not contain the alternating group of the set $\Gamma_1(x)$ and in view of theorem 2.4 - that $s \leqslant 3$. It follows from the result contained for example in the survey Weiss (1981 a) that either $G_1(x,y) = 1$ or $(G(x), \Gamma_1(x))$ is a normal extension of the group $PSL_n(q)$, $n \geqslant 3$ in its natural doubly transitive representation of degree $(q^n - 1)/(q - 1)$, $q = p^m$, p is a prime. In the latter case $| G_1(x,y) | = p^e$ for some integer e. To estimate the order of $G(x)$ in the case when $G_1(x,y) = 1$ we use the list Sims (1970) of primitive permutation groups having degree up to 20. Our notations mean the following: $G_1(x) = \left\{ g \mid g \in G(u) \text{ for all u such that } d(x,u) \leqslant i \right\}$, $G_1(x,y) = G_1(x) \cap G_1(y)$ for $\{x,y\} \in E(\Gamma)$.

If $a_1 \neq 0$ then we consider the graph Ξ such that $V(\Xi) = V(\Gamma) \cup V(\Delta)$ and vertices $x \in V(\Gamma)$, $y \in V(\Delta)$ are adjacent if the clique y contains the vertex x. The group $G = \mathrm{Aut}(\Gamma)$ acts naturally on Ξ and the orbits of G on $V(\Xi)$ are $V(\Gamma)$ and $V(\Delta)$. For a vertex $x \in V(\Xi)$ the stabilizer of x in G acts doubly transitively on the set $\Xi_1(x)$ of vertices adjacent to x. We shall make use of the following theorem by Weiss (1979).

<u>**Theorem 3.1.**</u> Let Ξ be a connected undirected graph, $\{x,y\} \in E(\Xi)$, $G \leqslant \mathrm{Aut}(\Xi)$ and $(G(x), \Xi_1(x))$, $(G(y), \Xi_1(y))$ are both primitive permutation groups. Let σ denote the set of primes dividing the order of $G_1(x,y)$. Then either $|\sigma| \leqslant 1$ or there is a number $p \in \sigma$ such that for $u = x$, $v = y$ or for $u = y$, $v = x$ the following holds: $O^p(G_1(x,y)) \leqslant G_2(u)$, $G_2(v)$ is a p-group and $G_2(v) = G_3(v)$.

Here $O^p(G_1(x,y))$ denotes the intersection of all normal subgroups of $G_1(x,y)$ having index p.

The application of theorem 3.1. shows that G acts either on the graph Γ or on the graph Δ in such a way that $G_1(x)$ is a p-group and $G_1(x)$ in its action on the set of vertices at distance 2 from x preserves as a whole every clique containing a vertex adjacent to x.

Notice that in theorem 3.1 we do not use the distance-transitivity of the graph Γ in question so the cases (k, a_1) and $((k/(a_1+1)-1)\cdot(a_1+2), (k/(a_1+1)-2))$ are dual via the consideration the clique graph.

Let us illustrate the theorem on the examples $(k, a_1) = (8, 1)$ and $(9, 2)$.

Let Γ be a graph of valency 8 such that $c_2 = a_1 = 1$. Then the subgraphs induced by the vertices which are at distance at most one from the vertices $x \in V(\Gamma)$ and $y \in V(\Delta)$ in the graphs Γ and Δ, respectively are presented on Fig. 1 and Fig. 2.

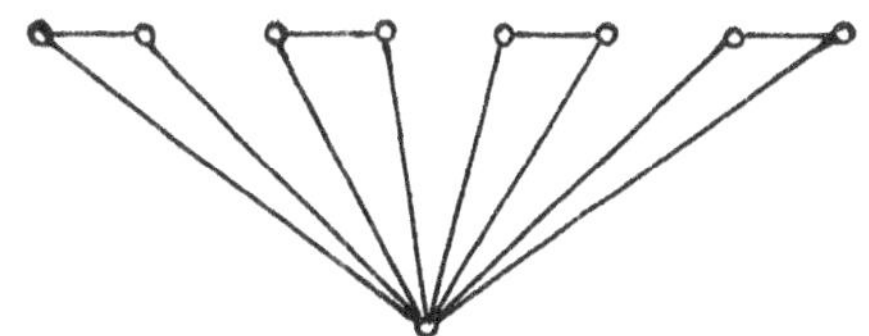

Fig. 1

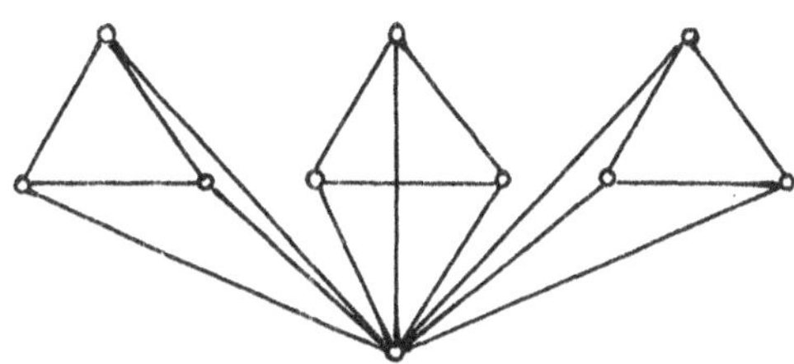

Fig. 2

It is clear that $\sigma \subseteq \{2,3\}$ and if $G_1(x,y)$ is a p-group then $|G(x)|$ divides $2^4 3 \cdot p^\infty$ for $p \in \{2,3\}$.

Let $|\sigma| > 1$. We consider the two possibilities

in theorem 3.1.

a. $u = y$, $v = x$, then $p = 2$ and $|G(x)|$ divides $2^\infty 3$;

b. $u = x$, $v = y$, then $p = 3$ and $|G(x)|$ divides $2^4 3^\infty$.

In the situation $k = 12$, $a_1 = 1$ an analogous arguments work.

Let $k = 10$, $a_1 = 1$ (or dually $k = 12$, $a_1 = 3$). Then for $x \in V(\Gamma)$, $y \in V(\Delta)$ the group $(G(x), \Xi_1(x))$ is one of the following: S_5, A_5, F_5^4 (F_5^4 is the Frobenius group of order 20) and $(G(y), \Xi_1(y))$ is the group S_3.

The situation $(G(x), \Xi_1(x)) = A_5$ or S_5 was investigated by Stellmacher (1984) and the situation $(G(x), \Xi_1(x)) = F_5^4$ by Fan (1986). From these papers we obtain the bounds presented in the Table 1.

Finally if $k = 12$, $a_1 = 2$ (a selfdual case) the groups $(G(x), \Xi_1(x))$ and $(G(y), \Xi_1(y))$ are both isomorphic to either A_4 or to S_4 and the main result of Stellmacher (1984 b) can be applied.

4 COMBINATORIAL CONDITIONS

This section contains some new conditions (feasibility conditions) necessary for the existence of d.t.g.'s. These conditions are formulated in terms of intersection arrays and can be applied to distance-regular graphs (d.r.g.'s) whose parameters satisfy the following relations:

$$(1, a_1, b_1) = \ldots = (1, a_{r-1}, b_{r-1}) \neq (1, a_r, b_r) \qquad (4.1)$$

The calculational experience shows that the analysis of intersection arrays satisfying (4.1) takes about 95% of the whole time of calculations.

It should be mentioned that the result of this section prove many particular cases of very significant conjecture 4.16 (see below).

For the sake of self-containment let us summarize briefly the main definitions and results of section 3 in Faradžev et al. (1986).

Let $x \in V(\Gamma)$ and $\{y_1, \ldots, y_k\} = \Gamma_1(x)$. For $u \in V(\Gamma)$, $u \neq x$ define the vector $\alpha^x(u) = (\alpha_1^x(u), \ldots, \alpha_k^x(u))$, where $\alpha_j^x(u) = d(u,y_j) - d(u,x)$, $j = 1, \ldots, k$, and $d(v,w)$ is the length of a shortest path connecting the vertices v and w.

Let $u \in \Gamma_s(x)$ and $\{z_1, \ldots, z_k\} = \Gamma_1(u)$. The matrix

$$A_s(u,x) = \begin{Vmatrix} \alpha^x(z_1) \\ \cdot \\ \cdot \\ \cdot \\ \alpha^x(z_k) \end{Vmatrix} = \begin{Vmatrix} \alpha_1^x(z_1) & \cdots & \alpha_k^x(z_1) \\ \cdot & & \cdot \\ \cdot & & \cdot \\ \cdot & & \cdot \\ \alpha_1^x(z_k) & \cdots & \alpha_k^x(z_k) \end{Vmatrix}$$

is called the characteristic matrix of the vertex u in respect to the vertex x.

The matrix $A_s(u,x)$ has a very convenient form if the neibours of the vertices u and x are ordered in such a way that the sequences $\{\alpha_i^x(u)\}_{i=1}^k$ and $\{\alpha_j^u(x)\}_{j=1}^k$ are nondecreasing. If so, the matrix $A_s(u,x)$ could be divided into blocks A_{nm}, $n, m = 1, 2, 3$. The possible values of elements in the blocks are indicated below:

$$A_s(u,x) = \begin{Vmatrix} A_{11} & A_{12} & A_{13} \\ A_{21} & A_{22} & A_{23} \\ A_{31} & A_{32} & A_{33} \end{Vmatrix} = \begin{Vmatrix} 0,\pm 1 & 0,\ 1 & 1 \\ 0,-1 & 0,\pm 1 & 0,\ 1 \\ -1 & 0,-1 & 0,\pm 1 \end{Vmatrix}$$

where A_{11}, A_{22}, A_{33} are square matrices of order c_s, a_s and b_s respectively.

The proofs of the following lemmas (4.1 - 4.5) can be found in Faradžev et al. (1986).

Lemma 4.1. Let $u \in \Gamma_s(x)$, $v \in \Gamma_{s+t}(x) \cap \Gamma_t(u)$, $t \geqslant 1$, then $\alpha_j^x(v) \leqslant \alpha_j^x(u)$ for all $1 \leqslant j \leqslant k$.

Lemma 4.2. If in the hypothesis of lemma 4.1 the equality $(c_s,\ a_s,\ b_s) = (c_{s+t},\ a_{s+t},\ b_{s+t})$ holds, then $\alpha^x(u) = \alpha^x(v)$.

Lemma 4.3. Let $u, v \in \Gamma_s(x)$, $d(u,v) = 1$. Then

i) if $(c_{s-1}, a_{s-1}, b_{s-1}) = (c_s, a_s, b_s)$ then the positions of -1's in the vectors $\alpha^x(u)$ and $\alpha^x(v)$ coincide;

ii) if $(c_{s+1}, a_{s+1}, b_{s+1}) = (c_s, a_s, b_s)$ then the positions of 1's in the vectors $\alpha^x(u)$ and $\alpha^x(v)$ coincide;

iii) if $(c_{s-1}, a_{s-1}, b_{s-1}) = (c_s, a_s, b_s) = (c_{s+1}, a_{s+1}, b_{s+1})$ then $\alpha^x(u) = \alpha^x(v)$.

Throughout this section we assume that the following holds for some $s \geqslant 1$, $t \geqslant 0$.

$$\left\| \begin{array}{c} c_{s-1} \\ a_{s-1} \\ b_{s-1} \end{array} \right\| \neq \left\| \begin{array}{c} c_s \\ a_s \\ b_s \end{array} \right\| = \ldots = \left\| \begin{array}{c} c_{s+t} \\ a_{s+t} \\ b_{s+t} \end{array} \right\| \neq \left\| \begin{array}{c} c_{s+t+1} \\ a_{s+t+1} \\ b_{s+t+1} \end{array} \right\| \quad (4.2)$$

Lemma 4.4. Let $x, z \in V(\Gamma)$, $d(x,z) = s-1$. Then modulo the replacement of x and z one of the following propositions holds:

i) $\alpha^x(w_1) \neq \alpha^x(w_2)$ for some $w_1, w_2 \in \Gamma_s(x) \cap \Gamma_1(z)$;

ii) $b_s = b_{s-1}$, $c_s \geqslant c_{s-1} + b_{s-1}$, $d < 2s-1$;

iii) $d = s$.

So the diameter of a d.r.g. is bounded by $2s-1$ if the conclusion i) of lemma 4.4 does not hold.

Let the conclusion i) of lemma 4.4 hold. Denote by d_s the minimal element of the set $\{ d(w_1, w_2) \mid w_1, w_2 \in \Gamma_s(x), \ \alpha^x(w_1) \neq \alpha^x(w_2) \}$. In view of lemma 4.4 the case $d_s \leqslant 2$ will be treated in detail.

In what follows we assume that $u, v \in \Gamma_s(x)$, $\alpha^x(u) \neq \alpha^x(v)$ and $d(u,v) = d_s$.

Lemma 4.5. If $d_s \leqslant 2$ then for some $1 \leqslant j \leqslant k$ $\alpha_j^x(u) = -1$, $\alpha_j^x(v) \geqslant 0$ holds

Let $w \in \Gamma_{s+q}(x) \cap \Gamma_q(v)$, $z \in \Gamma_{s+p}(x) \cap \Gamma_p(u)$, $p, q \geqslant 0$ and $W = (w_0, w_1, \ldots, w_l)$, $w_0 = z$, $w_l = w$ be a shortest path between the vertices w and z.

__Lemma 4.6.__ If $t \geqslant p + q + d_s - 1$, then $d(w,z) = p + q + d_s$.

__Proof.__ It follows from the construction that $d(w,z) \leqslant p + q + d_s$. By lemmas 4.2, 4.3 there are $0 \leqslant i,j \leqslant 1$ such that either w_i, $w_j \in \Gamma_s(x)$ or w_i, $w_j \in \Gamma_{s+t}(x)$. In the former case it is clear that $l \geqslant p + q + d_s$. Let us show that $l \geqslant p + q + d_s + 1$ in the latter case. Really, by lemmas 4.2, 4.3 i) and 4.5 $l \geqslant (t-p) + (t-q) + 2$. But $t \geqslant p + q + d_s - 1$, so $l \geqslant p + q + d_s + 1$.

If $c_s = 1$ and $z \in \Gamma_s(x)$, $s \geqslant 1$, the set $\Gamma_1(x) \cap \Gamma_{s-1}(z)$ consists of a unique vertex denoted below by $y(z)$. With the vertex z we associate the unique path $R(z) = (r_0, r_1, \ldots, r_s)$ such that $r_0 = x$, $r_1 = y(z)$, $r_s = z$. It is clear that $y(r_j) = y(z)$ for $1 \leqslant j \leqslant s$.

__Lemma 4.7.__ Let $u \in \Gamma_s(x)$, $c_s = 1$ and $V = \left\{ v_i \mid v_i \in \Gamma_s(x) \cap \cap \Gamma_1(u), y(v_i) \neq y(u) \right\}$. Then $|V| = a_s - a_{s-1}$. Moreover, if v_i, $v_j \in V$, $i \neq j$, then $y(v_i) \neq y(v_j)$.

__Proof.__ Let $\Gamma_1(u) \cap \Gamma_s(x) = \left\{ v_i \mid 1 \leqslant i \leqslant a_s \right\}$. Since $d(u,y(u)) = s-1$, we have $\left| \Gamma_1(u) \cap \Gamma_{s-1}(y(u)) \right| = a_{s-1}$. It follows from the equality $c_s = 1$ that $\Gamma_1(u) \cap \Gamma_{s-1}(y(u)) \subseteq \subseteq \Gamma_s(x)$. Without loss of generality we may assume that $y(v_i) = y(u)$ for $1 \leqslant i \leqslant a_{s-1}$ and $y(v_i) \neq y(u)$ for $a_{s-1} < i \leqslant \leqslant a_s$. Suppose that for $i,j > a_{s-1}$, $i \neq j$ and $y(v_i) = y(v_j) = y$. Then $d(u,y) = s$ and $c_s = \left| \Gamma_1(u) \cap \Gamma_{s-1}(y) \right| \geqslant \left| \{ v_i, v_j \} \right| = 2$, a contradiction with the equality $c_s = 1$.

Recall some definitions. A cycle $C = (x_0, x_1, \ldots, x_m)$, $x_0 = x_m$, $x_i \in V(\Gamma)$ is called undegenerate if for all $0 \leqslant i,j \leqslant m$ the distance between x_i and x_j in the cycle is equal to $d(x_i,x_j)$. Lemma 4.7 implies that if $u \in \Gamma_s(x)$, $c_s = 1$ then the path $R(u)$ is contained in exactly $(a_s - a_{s-1})$ undegenerate cycles of length $2s + 1$.

__Lemma 4.8.__ Let (4.1) and (4.2) hold. If $t \geq r$ then $b_s \leq k/2$.

__Proof.__ If $d_s > 2$ then it follows from lemma 4.4 that either $b_s = 0$ or $b_s = b_{s-1} \leq c_s - c_{s-1}$. In the latter case since $a_s + b_s + c_s = k$, we have $k - a_s - c_{s-1} \leq k-1$. Hence in both cases $b_s \leq k/2$. Therefore we assume that $d_s \leq 2$.

Let u and v be the same as above and let $q = r - d_s - 1$. Let $w \in \Gamma_{s+q}(x) \cap \Gamma_q(v)$ for $q \geq 0$ and $\{w\} = \Gamma_1(u) \cap \Gamma_1(v)$ if $q = -1$, i.e. if $d_s = r = 2$. Let $\mathcal{H}$ be the set of undegenerate cycles $H = (h_0, h_1, \ldots, h_{r-1}, h_r, h_{r+1}, \ldots, h_{2r+1} = h_0)$ of length $2r + 1$ such that $h_0 = u$, $h_{r-1} = w$, $h_r \in \Gamma_{s+q+1}(x)$. Since $c_r = 1$ the vertices h_i, $0 \leq i \leq r-1$ are the same in all cycles from $\mathcal{H}$ and by lemma 4.7 for a vertex $z \in \Gamma_{s+q+1}(x)$ there are exactly $(a_r - a_{r-1})$ cycles in $\mathcal{H}$ such that $h_r = z$. Hence

$$|\mathcal{H}| = \begin{cases} b_{s+q}(a_r - a_{r-1}) & \text{if } q \geq 0; \\ (b_{s-1} - 1)(a_r - a_{r-1}) & \text{if } q = -1. \end{cases} \quad (*)$$

In either case $|\mathcal{H}| \geq (b_s - 1)(a_r - a_{r-1})$.

Set $T = \Gamma_1(u) \setminus (\{h_1\} \cup \Gamma_{s+1}(x))$. We claim that for a cycle H contained in $\mathcal{H}$ it holds that $h_{2r} \in T$. Really, $h_{2r} \in \Gamma_1(u)$ by the definition of the cycle H. If $h_{2r} = h_1$ then the cycle is degenerate. Suppose that $h_{2r} \in \in \Gamma_{s+1}(x)$. Then by lemma 4.6 $d(h_r, h_{2r}) = r+1$, but since h_r and h_{2r} are in the cycle, $d(h_r, h_{2r}) = r$, a contradiction, and the claim is proved. From lemma 4.7 it follows that for a vertex $z \in T$ there are at most $(a_r - a_{r-1})$ cycles in $\mathcal{H}$ such that $h_{2r} = z$. If $d_s \leq 2$ then $h_1 \notin \Gamma_{s+1}(x)$, hence $|T| = a_s + c_s - 1$. Now we have

$$(a_s + c_s - 1)(a_r - a_{r-1}) \geq |\mathcal{H}| \geq (b_s - 1)(a_r - a_{r-1}).$$

So $b_s \leq a_s + c_s$ and $b_s \leq k/2$. This proves the lemma.

If (4.2) holds then in some cases the parameters of Γ enable to calculate d_s (see, for example lemma 3.10 in Faradžev et al.). In this situation lemma 4.8 can be made more precise.

__Lemma 4.9.__   i) if $d_s = 1$, $t \geqslant r$ or $t = r-1$ and $c_{s+t+1} = c_s$, then $2 \cdot b_s \leqslant k - c_{s-1} - 1$;

ii) if $d_s = 2$ and $t \geqslant r-1$ then

a) if $r > 2$ then $2 \cdot b_s \leqslant k - 1$ and if $2 \cdot b_s > k - c_{s-1} - 1$ then $(c_{s-2}, a_{s-2}, b_{s-2}) \neq (c_{s-1}, a_{s-1}, b_{s-1})$;

b) if $r = 2$ then $b_{s-1} \leqslant a_s + c_s$ and if $b_{s-1} > k - b_s - c_{s-1}$ then $(c_{s-2}, a_{s-2}, b_{s-2}) \neq (c_{s-1}, a_{s-1}, b_{s-1})$.

__Proof.__ Let us make all constructions described in the proof of lemma 4.8. Let $y_j \in \Gamma_1(x)$ be such a vertex that $\alpha_j^x(u) = -1$, $\alpha_j^x(v) \geqslant 0$. Let $d_s = 1$ and either $t \geqslant r$ or $t = r-1$ and $c_{s+t+1} = c_s$. Set $F = \Gamma_1(u) \cap \Gamma_{s-2}(y_j)$, $|F| = c_{s-1}$. If $h_{2r} \in F$ for some cycle H in $\mathcal{H}$ then an application of lemmas 4.1 - 4.3 to all possible configurations of the cycle gives $\alpha_j^x(v) = -1$. Hence $h_{2r} \in T \setminus F$. Meanwhile, since $q \geqslant 0$, $|\mathcal{H}| = b_{s+q} \cdot (a_r - a_{r-1})$, so $2 \cdot b_s \leqslant k - c_{s-1} - 1$.

If $d_s = 2$ then all arguments of lemma 4.8 are valid even when $t = r-1$. Moreover, if $2 \cdot b_s > k - c_{s-1} - 1$ for $r > 2$ and $b_{s-1} > k - b_s - c_{s-1}$ for $r = 2$ then there is a cycle $H \in \mathcal{H}$ such that $h_{2r} \in F$. If $h_{2r-1} \in \Gamma_s(x)$ then lemmas 4.1 - 4.3 being applied to all possible configurations of the cycle H imply that $\alpha_j^x(v) = -1$, this is a contradiction. If $h_{2r-1} \in \Gamma_{s-1}(x)$ then $\alpha_j^x(h_{2r-1}) = \alpha_j^x(w) = \alpha_j^x(v) \neq \alpha_j^x(u) = \alpha_j^x(h_{2r})$. Since $d(h_{2r}, h_{2r-1}) = 1$, lemma 4.3 i) implies that $(c_{s-2}, a_{s-2}, b_{s-2}) \neq (c_{s-1}, a_{s-1}, b_{s-1})$ and the lemma is proved completely.

Notice that the following lemma is also true.

__Lemma 4.10.__ Let the relations (4.1) - (4.2) hold and $t \geqslant r$. Then $c_s \leqslant k/2$.

To prove the lemma one should prove a row of propositions analogous to lemmas 4.4 - 4.6. In this case the relations (4.2) should be considered from right to left instead of consideration from left to right carried out in the previous case.

Now we proceed to a detailed consideration of the relations (4.2) under the additional assumption $c_s = 1$. One of the main facts concerning this situation is contained in the following

Lemma 4.11. Let $z \in \Gamma_s(x)$, $c_s = 1$ and $y \in \Gamma_1(x) \smallsetminus \{y(z)\}$. Set $T(y) = \{v \mid 1 \leqslant d(x,v) \leqslant s, \ y(v) = y, \ \Gamma_1(v) \cap R(z) \neq \emptyset\}$. Then $|T(y)| \leqslant 1$.

Proof. Suppose that $v_1, v_2 \in T(y)$, $v_1 \neq v_2$. Let $u_i \in \Gamma_1(v_i) \cap R(z)$ for $i = 1,2$. If $u_1 = u_2 = r_j$ for some $1 \leqslant j \leqslant s$ then $d(r_j, y) = j$ and the set $\Gamma_1(r_j) \cap \Gamma_{j-1}(y)$ contains the at least two vertices, namely v_1 and v_2 - a contradiction with the equality $c_j = 1$. Therefore we assume that $u_1 = r_1$, $u_2 = r_j$ for $1 < j$. Since $y(v_i) = y \neq y(u_i)$ and $c_s = 1$, the lemmas 4.2 - 4.3 imply that $d(x,v_i) = d(x,u_i)$ for $i = 1,2$. For the vertex $r_{j-1} \in \Gamma_1(r_j) \cap \Gamma_{j-1}(x)$ one has $d(r_{j-1}, y) = d(r_{j-1}, r_1) + d(r_1, v_1) + d(v_1, y) = j-1$. Hence $r_{j-1}, v_2 \in \Gamma_1(r_j) \cap \Gamma_{j-1}(y)$, again a contradiction with $c_j = 1$.

The following lemma firstly appeared in Faradžev et al. (1986), lemma 3.11.

Lemma 4.12. Let the relations (4.1) - (4.2) hold. Then

i) if $c_s = 1$ then $t \leqslant r-1$, moreover, if $t = r-1$ then $c_{s+t+1} > 1$;

ii) if $b_s = 1$ and $d > s+t+1$ then $t \leqslant r-1$, moreover, if $t = r-1$ then $b_{s-1} > 1$.

The following propositions strengthen the conclusion ii) of lemma 4.12.

Lemma 4.13. Let the relations (4.1) - (4.2) hold, $c_s = 1$ and $r \geqslant 3$. Then at least one of the following hold

i) $t < r-2$;

ii) $c_{s+r-1} \geqslant 2$;

iii) $(a_s - a_{s-1}) \geqslant (a_r - a_1 + 1)$.

__Proof.__ If in the relations (4.2) the equality $c_s = 1$ holds
then a consideration of the matrix $A_s(u,x)$ shows that
$d_s = 1$ (see Faradžev et al. (1986) for the details). Let
$W = (w_0, w_1, \ldots, w_{r-1})$ be the path such that $w_0 = u$, $w_1 = v$ and $w_i \in \Gamma_{s+i-1}(x)$ for $2 \leqslant i \leqslant r-1$. If $t \geqslant r-2$ then $c_{s+r-2} = c_s = 1$ and hence $d(u, w_{r-1}) = r-1$. In this case lemma 4.7
implies that the vertex w_{r-1} has $m = a_r - a_1$ neighbours
(say $z_1, z_2, \ldots, z_m$) which are at distance r from the vertex $u_0 \in \Gamma_1(u) \cap \Gamma_{s-1}(x)$. If $z_i \in \Gamma_{s+r-1}(x)$ for some
$1 \leqslant i \leqslant m$ then $c_{s+r-1} \geqslant 2$ and the conclusion ii) is valid.
Thus if $t > r-2$ and $c_{s+r-1} = 1$ then $z_i \in \Gamma_{s+r-2}(x)$ for
$1 \leqslant i \leqslant m$. By lemmas 4.2, 4.3 in this case $y(z_i) = y(w_{r-1}) = y(v)$. Let us consider the shortest path $Q(z_i) = (q_0^i, q_1^i, \ldots, q_r^i)$ from the vertex $z_i = q_0^i$ to the vertex $u_0 = q_r^i$. The
following conditions hold for all vertices q_j^i contained in
the paths $\{Q(z_i) \mid 1 \leqslant i \leqslant m, \ 0 \leqslant j \leqslant r\}$.

a) $d(w_{r-1}, q_j^i) = j+1$ for $1 \leqslant i \leqslant m$, $0 \leqslant j \leqslant r-1$;
b) $q_j^i \neq q_g^f$ if $i \neq f$ or $j \neq g$.

If the condition a) fails then Γ contains a
cycle of lenght less then $2 \cdot r + 1$ which is not a triangle.
If the condition b) fails then $c_i \geqslant 2$ for some $i \leqslant r$.
It is clear that the path $Q(z_i)$ contains
exactly one pair of vertices $\{q_j^i, q_{j+1}^i\}$ at the same
distance from the vertex x. This and lemmas 4.2 - 4.3 give
$y(q_j^i) = y(z_i)$, $y(q_{j+1}^i) = y(u_0)$ and either $j = r-2$ or $j = r-1$. Moreover, the second possibility contradicts to lemma 4.11. This means that a vertex $u_0 \in \Gamma_{s-1}(x)$ has in
$\Gamma_s(x)$ $(a_r - a_1 + 1)$ neighbours $\{q_{r-1}^i \mid 1 \leqslant i \leqslant m\} \cup \{u\}$
which are at distance s from $y(v)$. Hence we have the
inequality $(a_s - a_{s-1}) \geqslant (a_r - a_1 + 1)$.

Let r be the minimal number such that
$(1, a_1, b_1) \neq (c_r, a_r, b_r)$. In this case the length g of
a shortest undegenerate cycle which is different from a
triangle is equal to $2 \cdot r$ if $c_r \neq 1$ and to $2 \cdot r + 1$ otherwise.
Notice that the value g coincides with the girth of Γ if

$a_1 = 0.$

The following lemma was proved by Terwilliger (1983).

<u>Lemma 4.14.</u> Let Γ be a distance-regular graph with the girth g equal to $2 \cdot 1$ or $2 \cdot 1 - 1$ and $a_{1-1} \leqslant 2 \cdot c_1 - 2.$ Then

$$\frac{p}{c_1 - 1} \geqslant \frac{b_i}{b_{i-1+1} - 1} + \frac{c_{j-1+1}}{c_j - 1} \qquad (4.3)$$

where $p = \max \{c_1, a_{1-1}\}.$

Let Γ be a d.r.g. satisfying the conditions of lemma 4.14 and having the girth g. Then $r = \left[g/2 \right]$, in particular $1 \leqslant r + 1.$

It is easy to prove that if $a_{1-1} \leqslant 2 \cdot c_1 - 2,$ the inequality (4.3) implies that $t \leqslant r.$ Really, let $t \geqslant r+1$ and $(c_s, a_s, b_s) = (c_{s+t}, a_{s+t}, b_{s+t}).$ Putting in (4.3) $j = i+1 = s+t,$ we obtain

$$2 \geqslant \frac{p}{c_1 - 1} \geqslant \frac{b_{s+t-1}}{b_{s+t-1} - 1} + \frac{c_{s+t-1+1}}{c_{s+t} - 1} \geqslant 1 + \frac{1}{b_s - 1} + 1 + \frac{1}{c_s - 1}$$

this is a contradiction.

This result of P.Terwilliger can be extended to the case $a_1 \neq 0.$ Namely the following lemma is true.

<u>Lemma 4.15.</u> Let Γ be a distance-regular graph. Let g be the length of a shortest undegenerate cycle which is different from a triangle. For $g = 2 \cdot 1 - 1$ let $a_{1-1} \leqslant 2 \cdot c_1 - 2.$ Then the inequality (4.3) holds, where $p = c_1$ for $g = 2 \cdot 1$ and $p = \max \{c_1, a_{1-1}\}$ for $g = 2 \cdot 1 - 1.$

<u>Proof</u> coincides with that of lemma 4.14 modulo the substitution of the term "the shortest cycle" by the term "the shortest undegenerate cycle different from a triangle".

Notice that lemmas 4.8, 4.10, 4.12 and 4.15 contain the proof of some particular cases of the following very significant conjecture.

<u>Conjecture 4.16.</u> Let Γ be a distance-regular graph and the inequalities (4.1), (4.2) hold for some t. Then $t \leqslant r$.

Notice the following lemma proved by Gardiner (1981).

<u>Lemma 4.17.</u> Let $a_1 = a_2 = \ldots = a_{r-1} = 0$, $a_r \neq 0$. Then the set $\{ i \mid a_i = 0, \; s \leqslant i \leqslant s + t \}$ is of cardinality at most r.

By lemmas 4.8, 4.10, 4.12, 4.15 and 4.17 the conjecture 4.16 can be false only if $2 \leqslant b_s \leqslant k/2$, $2 \leqslant c_s \leqslant k/2$, $a_s \geqslant 1$.

5 <u>SOME GENERAL PROPERTIES OF IMPRIMITIVE GRAPHS</u>

In the analysis of intersection arrays obtained by the algorithm for $8 \leqslant k \leqslant 13$ the case of imprimitive graphs was the most difficult. We begin with the antipodal graphs. In all cases the identification of the antipodal quotient was quite easy. So the construction of antipodal graphs was reduced to the construction of distance-transitive antipodal coverings of given d.t.g.'s. In this section a row of propositions concerning this construction are presented. These propositions are useful both in construction of graphs and in proving of their nonexistence. A basis of the theory of antipodal d.t.g.'s can be found in Gardiner (1974 a,b).

Let Γ be an antipodal d.t.g. of diameter $d \geqslant 3$ and of valency $k \geqslant 3$. Let $G \leqslant \text{Aut}(\Gamma)$ act distance-transitively on Γ, $\bar{\Gamma}$ be the antipodal quotient of Γ and N be the kernel of the action of G on $\bar{\Gamma}$. In other words N is the largest subgroup in G, preserving every antipodal block of Γ as a whole. Let $\mathcal{B} = \{ B_i \mid i \in J \}$ be the set of antipodal blocks of Γ, then $B_i = \{x\} \cup \Gamma_d(x)$ for a ver-

tex $x \in B_1$. The graph Γ is called l-fold antipodal cove-
ring of the graph $\bar{\Gamma}$, $l = |B_1|$, $B_1 \in \mathcal{B}$. The factor-
group G/N will be denoted by $\bar{G}$.

We begin with the case when $\bar{G} \cong G$, i.e. when
$N = 1$. In this case $G(x)$ is a subgroup of $G(\bar{x})$ where $\bar{x} = \{x\} \cup \Gamma_d(x)$ is the antipodal block containing x.

Lemma 5.1. If $N = 1$ then

i) the permutational representation of $G(\bar{x})$ on
the set of cosets of $G(x)$ is doubly transitive;

ii) $G(x)$ as the subgroup of $G(\bar{x})$ acts transitively on the set $\bar{\Gamma}_i(\bar{x})$ for all $0 \leq i \leq [d/2]$.

iii) if $G(x) < H < G$ then either $H = G(\bar{x})$ or
$[G:H] = 2$ and the graph Γ is bipartite.

Proof. i) Let $\bar{x} = \{x\} \cup \Gamma_d(x)$. Since $\bar{x}$ is an imprimitivity
block of G, the subgroup $G(\bar{x})$ acts transitively on this set.
On the other hand $G(x)$ is the stabilizer of a vertex x in
this action and $G(x)$ acts transitively on the set $\Gamma_d(x)$.
ii) The group G acts distance-transitively on the graph Γ
and $G(x)$ acts transitively on the set $\Gamma_i(x)$ for $0 \leq i \leq d$.
This means that $G(x)$ acts transitively on the set of antipodal blocks of Γ having nonempty intersection with the
set $\Gamma_i(x)$ and the claim follows. iii) This conclusion follows from the fact that a distance-transitive graph has at
most two imprimitivity systems - a bipartite system and an
antipodal system.

Now let $N \neq 1$.

Lemma 5.2. N is semiregular on $V(\Gamma)$.

Proof. Suppose that an element $n \in N$ fixes a vertex $x \in V(\Gamma)$.
Let $x \in B$, $B \in \mathcal{B}$ and B_i, $1 \leq i \leq k$ be the blocks containing the
vertices adjacent to x. Since $d \geq 3$ the vertex x is adjacent
to exactly one vertex in the block B_i, $1 \leq i \leq k$. On the
other hand the element n stabilizes every antipodal block.
This implies that n fixes the vertices in the set $\Gamma_1(x)$.

Since Γ is connected it is easy to show that $n = 1$.

Lemma 5.3. N acts regulary and faithfully on the vertices in a block $B \in \mathcal{B}$.

Proof. From lemma 5.2 it follows that the action of N on B is faithfull. Since $N \triangleleft G$ the orbits of N are the imprimitivity blocks of the group G. On the other hand $\mathcal{B}$ is the minimal imprimitivity system, hence $\mathcal{B}$ consists of the orbits of N on $V(\Gamma)$.

Lemma 5.4. N is an elementary abelian p-group for some prime number p.

Proof. Let us consider the stabilizer H=H(B) of a block $B \in \mathcal{B}$ as a whole. Since $B = \{x\} \cup \Gamma_d(x)$, the group G acts distance transitively on Γ and B is an imprimitivity block, the subgroup H acts doubly transitively on B   and   N is an elementary abelian p-group for some prime number p.

Let $|N| = p^n$. The next lemma is a direct consequence of lemmas 5.3 and 5.4.

Lemma 5.5. There is a homomorphism φ of the group G into the group $GL_n(p)$ such that the immage of φ acts transitively on the set of nonzero vectors of the n-dimensional space over $GF(q)$.

Lemma 5.6. If $G(x) < H < G$ then either $H = \langle G(x), N \rangle$, or $[G{:}H] = 2$ and Γ is a bipartite graph.

Proof follows from the description of possible imprimitivity systems of the graph Γ .

The group G is an extension of the group N by the group $\bar{G}$ ($G = N.\bar{G}$). In many cases lemma 5.6 enable to show that the extension is nonsplit, i.e. G is not isomorphic to a semidirect product of N and $\bar{G}$.

Now let us consider the bipartite graphs of diameter three. Such a graph has the following intersection array:

$$\{k,\ k-1,\ k-\lambda\ ;\ 1,\ \lambda\ ,\ k\}$$

It is well known that if we take the vertices in one partition as the points and in other partition as the blocks with the natural incidence relation, we obtain a symmetric $2-(v, k, \lambda)$ - design. Here v is the halved number of the vertices in Γ . If Γ is a d.t.g. then the subgroup of Aut(Γ), preserving everyhalf acts doubly transitively both on the set of points and on the set of blocks. In addition the stabilizer of a block acts transitively on the set of points contained in the block.

Modulo the completeness of the list of known doubly transitive permutation groups a design possessing the above properties is either one of the following or the complementory design to one of the following:

i) the degenerated design; v points and v blocks, each point is incident to each block (Γ is the complete bipartite graph $K_{v,v}$);

ii) the design consisting of v blocks, each of cardinality v-1 (Γ is the complete bipartite graph $K_{v,v}$ without a matching);

iii) the design, whose points are lines and blocks are hyperplanes of the n-dimensional space over the field with q elements: $v = (q^n - 1)/(q - 1)$, $k = (q^{n-1} - 1)/(q - 1)$, $\lambda = (q^{n-2} - 1)/(q - 1)$;

iv) the symplectic designs (see Kantor (1975)) $v = 2^{2m}$, $k = 2^{m-1}(2^m + \varepsilon)$, $\lambda = 2^{m-1}(2^{m-1} + \varepsilon)$, $m \geqslant 2$, $\varepsilon = \pm 1$;

v) the Paley design with 11 points: $v = 11$, $k = 5$, $\lambda = 2$;

vi) the Higman design with 176 points (see Higman(1969)): $v = 176$, $k = 50$, $\lambda = 14$.

6 A DISCUSSION OF THE RESULTS

Now we proceed to discussion of the constructed intersection arrays and of the graphs with these arrays. As usual we begin with infinite series. We certainly have the series of graphs which are characterized in section 2 and in section 5 (the graphs related to the symmetric designs). Besides that the following series occur.

1. The complete multipartite graphs. These graphs can be characterized by the property that they are antipodal of diameter 2.

2. The Johnson graphs $J(n,1)$ and the complementary graphs of the graphs $J(n,2)$.

3. The Hamming graphs $H(n,m)$, the complementary graphs of $H(2,m)$ and the folded graphs of $H(n,2)$. Notice that $H(n,2)$ is the n-dimensional cube, characterized in lemma 2.1.

4. The point graphs and the line graphs of the classical generalized n-gons.

It is well known (see, for example Roos (1980)) that the point graph and the line graph of a generalized n-gon are diatance-regular of diameter $[n/2]$ having the intersection arrays

$$\{t(s + 1),\ st,\ \ldots,\ st;\ 1,\ 1,\ \ldots,\ s + 1\} \quad \text{and}$$
$$\{s(t + 1),\ st,\ \ldots,\ st;\ 1,\ 1,\ \ldots,\ t + 1\} \quad ,$$

respectively.

The generalized n-gons arising from groups having a (B,N)-pair of rank 2 are called classical. In the case of classical generalized n-gons the point graph and the line graph are d.t.g.'s.

The list of known groups with (B,N)-pair of rank 2 is presented in Table 2, taken from the paper Higman (1974).

<u>TABLE 2</u>

Type	Identification	W	n	s	t
$A_2(q)$	$PSL_3(q)$	D_6	3	q	q
$B_2(q)$	$PSp_4(q)$	D_8	4	q	q
$^2A_3(q)$	$PSU_4(q)$	D_8	4	q^2	q
$^2A_4(q)$	$PSU_5(q)$	D_8	4	q^2	q^3
$G_2(q)$	Dickson's group	D_{12}	6	q	q
$^3D_4(q)$	triality group	D_{12}	6	q^3	q
$^2F_4(q),\ q=2^{2m+1}$	Ree group	D_{16}	8	q^2	q

It is clear that if $s = t$ then the parameters
of the point graph and of the line graph coincide. These
graphs are isomorphic for all q in the case of 3-gons of
type $A_2(q)$. For n-gons of type $B_2(q)$ and $G_2(q)$ these graphs
are isomorphic if and only if $q = 2^m$ and $q = 3^m$, respectively.

5. The graphs $r.K_{k,k}$ — r-fold antipodal coverings of the graphs $K_{k,k}$. It was shown in Faradžev et al.
(1986) that there is a correspondence between these graphs
and resolvable transversal designs $R[k, k/r, r]$ in the
notations of Hanani (1974).

If $r = k$ the designs are related to projective
planes of order k. For the values k in question the methods
presented in the section 5 enable to show that these planes
are desargesian.

If $r = 2$ the design are related to Hadamard
matrices of order k. Hence, in particular, $k = 0 \pmod 4$.
This implies for example the nonexistence of the graph

$2.K_{10,10}$. For k = 8 and k = 12 we have the Hadamard matrices comming from the Paley designs of the field GF(q) for q = 7 and q = 11, respectively.

If $2 < r < k$ we have two examples - the graph $4.K_{8,8}$ and $3.K_{9,9}$ which are related to the designs R[8, 2, 4 and R[9, 3, 3] , respectively, described in Hanani (1974).

6. The q-analogs of the double covering of the odd graphs. For the values k in question we have one example. This is the graph with the intersection array

$$\{13, \ 12, \ 12, \ 9, \ 9; \ 1, \ 1, \ 4, \ 4, \ 13\}$$

and with the automorphism group isomorphic to the extension of the group $PGL_5(3)$ by the contragredient automorphism.

7. The Paley graphs over the field with q elements, q = 1(mod 4). These graphs have valency k = (q - 1)/2.

8. Double antipodal coverings of the complete graphs K_{k+1} such that the subgraph induced on the neighbourhood of a vertex is the Paley graph. Notice that for q = 9 we have the Johnson graph J(6,3)

In the Table 3 the d.t.g.'s of valency $8 \leqslant k \leqslant 13$ which are not members of the above mentioned series are presented. In the fourth column of the table it is indicated whether the graph is primitive (p.) or antipodal and/or bipartite (a. and/or b.).

It is shown by Brouwer (1984) that there are exactly two graphs with the intersection array № 1 in the Table 3. These graphs are related to the spreads of the point graph of the generalized 4-gon of type $^2A_3(2)$. It can be proved that only one of these graphs is d.t.g.

Finally in Table 4 we give the list of intersection arrays for which it was shown by individual arguments that no d.t.g.'s exist. In the nonexistence proofs we have used in particular the list of primitive graphs having at most 50 vertices, contained in Faradžev & Klin (1985).

TABLE 3

N°	$|V(\Gamma)|$	$i(\Gamma)$		Aut(Γ)	References
1.	27	$\{8,6,1;1,3,8\}$	a.	$Z_3.E_9.GL_2(3)$	Brouwer (1984)
2.	280	$\{9,8,6,3;1,1,3,8\}$	p.	$P\Sigma L_3(4)^*$	Biggs (1976), Tchuda (1985)
3.	63	$\{10,6,4,1;1,2,6,10\}$	a.	$Z_3.S_7$	Smith (1977)
4.	65	$\{10,6,4;1,2,5\}$	p.	$P\Sigma L_2(25)$	A.A.Ivanov (1981), Gordon & Levingstone (1981)
5.	315	$\{10,8,8,2;1,1,4,5\}$	p.	Aut(J_2)	Cohen(1981), Zaytchenko (1981)
6.	56	$\{10,9;1,2\}$	p.	$P\Sigma L_3(4)^*$	Gewirtz graph
7.	112	$\{10,9,8,2,1;1,2,8,9,10\}$	b.a.	$P\Sigma L_3(4)^* \times Z_2$	
8.	266	$\{11,10,6,1;1,1,5,11\}$	p.	J_1	Biggs (1976)
9.	68	$\{12,10,3;1,3,8\}$	p.	$P\Gamma L_2(16)$	Zaytchenko (1981), Gordon & Levingstone (1981)
10.	208	$\{12,10,5;1,1,8\}$	p.	$P\Gamma U_3(4)$	Zaytchenko et al. (1980), Gordon & Levingstone (1981)

TABLE 4

№	$\|V(\Gamma)\|$	$i(\Gamma)$	
1.	63	$\{8,6,1,;1,1,8\}$	a.
2.	100	$\{8,7,6,4;1,2,4,8\}$	b.
3.	40	$\{9,6,1;1,2,9\}$	a.
4.	650	$\{9,8,8,7,1;1,1,2,8,9\}$	b.
5.	21	$\{10,5;1,5\}$	p.
6.	26	$\{10,6;1,4\}$	p.
7.	33	$\{10,6,1;1,3,10\}$	a.
8.	99	$\{10,8,1;1,1,10\}$	a.
9.	112	$\{10,9,1,1;1,1,9,10\}$	a.
10.	112	$\{10,9,4,2,1;1,2,4,9,10\}$	a.
11.	100	$\{10,9,8,1;1,2,9,10\}$	b.a.
12.	164	$\{10,9,8,5;1,2,5,10\}$	b.
13.	200	$\{10,9,8,6;1,2,4,10\}$	b.
14.	210	$\{11,10,4,1,1,5\}$	p.
15.	88	$\{12,10,2;1,2,8\}$	p.
16.	143	$\{12,10,1;1,1,12\}$	a.
17.	72	$\{12,11,8,1;1,4,11,12\}$	b.a.
18.	96	$\{12,11,9,1;1,3,11,12\}$	b.a.
19.	180	$\{12,11,9,6,3,1;1,3,6,9,11,12\}$	b.a.
20.	222	$\{12,11,9,8;1,3,4,12\}$	b.
21.	270	$\{12,11,9,8,3,1;1,3,4,9,11,12\}$	b.a.
22.	144	$\{12,11,10,1;1,2,11,12\}$	b.a.
23.	244	$\{12,11,10,6;1,2,6,12\}$	b.
24.	288	$\{12,11,10,7;1,2,5,12\}$	b.
25.	42	$\{13,8,1;1,4,13\}$	a.
26.	442	$\{13,12,12,4;1,1,9,13\}$	b.

REFERENCES

1. Armanios, C. (1981). Symmetric graphs and their automorphism groups. Ph. D. Thesis, Univ. of
   Western Australia, Perth.
2. Bannai,E. & Ito, T. (1984). Algebraic Combinatorics I.
   Association Schemes. Benjamin.:California.
3. Biggs,N.L. (1974). Algebraic Graph Theory. Cambridge.:
   Cambridge Univ. Press.
4. Biggs,N.L. (1974). The symmetry of line graphs. Util.
   Math.,5, 113-121.
5. Biggs,N.L. (1976). Automorphic graphs and the Krein
   condition. Geom. Dedic., 5, 117-127.
6. Biggs,N.L. & Smith,D.H. (1971). On trivalent graphs.
   Bull. London Math. Soc., 3, 155-158.
7. Brouwer, A.E. (1983). On the uniqueness of a regular
   thin near octagon on 288 vertices (or the semi-
   biplane belonging to the Mathieu group M_{12}).
   Math. Cent. Afd. zuivere wisk № 196, 12pp.
8. Brouwer, A.E. (1984). Distance-regular graphs of diame-
   ters 3 and strongly regular graphs. Discrete
   Math., 49, 101-103.
9. Brouwer, A.E., Cohen,A.M. & Neumaier, A. (1987).Distan-
   ce Regular Graphs. A preliminary version of a
   book.
10. Buekenhout, F. & Rowlinson, P. (1981). The uniqueness
    of certain automorphic graphs. Geom. Dedic.,
    11, 443-446.
11. Cameron, P.J. (1974). Suborbits in transitive permuta-
    tion groups. In Combinatorial Group Theory.
    Math. Centre Tracts № 57, pp. 98-129. Amster-
    dam: D. Reidel.
12. Cameron, P.J. (1981). Finite permutation groups and
    finite simple groups. Bull. London Math. Soc.,
    13, 1-22.
13. Cameron, P.J. (1982). There are only finitely many dis-
    tance-transitive graphs of given valency grea-
    ter than two. Combinatorica, 2, 9-13.

14. Cameron,P.J., Praeger, C.E., Saxl, J. & Seitz, G.M.
         (1983). On the Sims conjecture and distance-
         transitive graphs. Bull. London Math. Soc.,
         15, 499-506.
15. Cohen, A.M. (1981). Geometries originating from certain
         distance-regular graphs. London Math. Soc.
         Lect. Note Ser., 49, 81-87.
16. Fan, P.S. (1986). Amalgams of prime index.J. Algebra,
         98, 375-421.
17. Faradžev, I.A., Ivanov,A.A. & Ivanov, A.V. (1986). Dis-
         tance-transitive graphs of valency 5, 6 and 7.
         Europ. J. Comb., 7, 303-319.
18. Gardiner, A. (1973). Arc transitivity in graphs. Quart.
         J. Math., 24, 399-407.
19. Gardiner, A. (1974). Antipodal covering graphs. J. Comb.
         Theory (B), 16, 255-273.
20. Gardiner, A. (1974). Imprimitive distance-regular graphs
         and projective planes. J. Comb. Theory (B), 16,
         274-281.
21. Gardiner,A. (1975). On trivalent graphs. J. London.
         Math. Soc., 10, 507-512.
22. Gardiner, A. (1981). When is an array realized by a
         distance-regular graph? In Algebraic Methods
         in Graph Theory, vol. 1, pp. 209-219. Amster-
         dam: Horth-Holland.
23. Gardiner, A. (1982). Classifying distance-transitive
         graphs. Lecture Notes Math. 952, 67-88.
24. Gardiner, A. (1985). An elementary classification of
         distance-transitive graphs of valency four.
         Ars. Comb., 19A, 129-141.
25. Gardiner, A. & Praeger, C.E. (1986). Distance-transitive
         graphs of valency six. Ars. Comb., 21A, 195-
         210.
26. Gardiner, A. & Praeger, C.E. (1987). Distance-transitive
         graphs of valency five. Proc. Edinburgh Math.
         Soc., 30, 73-81.
27. Gleason, A.M. (1956). Finite Fano planes. Amer. J. Math.
         78, 797-807.

28. Goldschmidt, D. (1980).Automorphisms of trivalent
        graphs. Ann. Math., 111, 377-406.
29. Gordon, L.M. & Levingstone, R. (1981). The construction
        of some automorphic graphs. Geom. Dedic., 10,
        261-267.
30. Hanani, H. (1974). On transversal designs. In Combinato-
        rial Group Theory. Math. Centre Tracts № 57,
        pp. 43-53. Amsterdam: D. Reidel.
31. Higman, D.G. (1964). Finite permutation groups of rank
        3. Math. Z., 86, 145-156.
32. Higman, D.G. (1974).Invariant relations, coherent confi-
        gurations and generalized polygons. In Combina-
        torial Group Theory. Math. Centre Tracts № 57,
        pp. 347-364. Amsterdam: D. Reidel.
33. Higman, G. (1969). On the simple group of D.G.Higman
        and C.C.Sims. Illinois J. Math., 13, 74-84.
34. Ivanov, A.A. (1981). Construction by the computer of
        some new automorphic graphs. In Aerofizika i
        prikladnaia matematika. pp. 144-146. Moscow:
        MFTI. (In Russian)
35. Ivanov, A.A. (1983). Bounding the diameter of a distan-
        ce-regular graph. Soviet Math. Dokl. 28, 149-
        152.
36. Ivanov, A.A. (1984). Combinatorial and algebraic methods
        for investigation of distance-regular graphs.
        Ph. D. Thesis. Moscow. MFTI. (In Russian)
37. Ivanov, A.A. (1986). Graphs of small girth and distance-
        transitive graphs. In Metodi issledovaniya
        slozhnih sistem. pp. 13-19. Moscow: Institute
        for System Studies. (In Russian)
38. Ivanov, A.A. & Chuvaeva, I.V. (1985). Fixed points me-
        thod in classification of distance-transitive
        graphs. In Metodi issledovaniya slozhnih sistem
        pp. 3-10. Moscow: Institute for System Studies.
        (In Russian)
39. Kalužnin, L.A., Suschansky, V.I. & Ustimenko, V.A.
        (1982). The use of computer in the theory of
        permutation groups and their applications.

Kibernetika, $\underline{6}$, 83-94. (In Russian)

40. Kantor, W.M. (1975). Symplectic groups, symmetric designs and line ovals. J. Algebra, $\underline{33}$, 43-58.

41. Klin, M.H. & Faradžev, I.A. (1985). The primitive graphs with fever then 50 vertices. Preprint. (In Russian)

42. Praeger, C.E. (1980). Symmetric graphs and a characterization of the odd graphs. Lecture Notes Math., $\underline{829}$, 211-219.

43. Quirin, W.L. (1971).Primitive permutation groups with small orbitals. Math. Z., $\underline{122}$, 267-274.

44. Roos, C. (1980). An alternative proof of Feit - Higman theorem. Delft. Prog. Rept. $\underline{5}$, 67-77.

45. Sims, C.C. (1967). Graphs and finite permutation groups I. Math. Z., $\underline{95}$, 76-86.

46. Sims, C.C. (1968). Graphs and finite permutation groups II. Math. Z., $\underline{103}$, 276-281.

47. Sims, C.C. (1970). Computation methods in the study of permutation groups. In Computation Problems in Abstract Algebra, pp. 169-184. Pergamon Press.

48. Smith, D.H. (1973). On tetravalent graphs. J. London Math. Soc., $\underline{6}$, 659-662.

49. Smith, D.H. (1974). Distance-transitive graphs of valency four. J. London Math. Soc., $\underline{8}$, 377-384.

50. Smith, D.H. (1974). On bipartite tetravalent graphs. Discrete Math., $\underline{10}$, 167-172.

51. Smith, S.D. (1977). Nonassociative commutative algebras for triple covers of 3-transposition groups. Michigan Math. J., $\underline{24}$, 273-287.

52. Stellmacher, B. (1984). On graphs with edge-transitive automorphism group. Illinois J. Math., $\underline{28}$, 211-266.

53. Stellmacher, B. (1984). Rank 2 groups. In Proceedings of the Rutgers Group Theory Year, 1983-1984. pp. 197-210. Cambridge: Cambridge Univ. Press.

54. Tchuda, F.L. (1985). Construction of an automorphic graph on 280 vertices using finite geometries. In Investigation on Algebraic Theory of Combi-

natorial Objects. pp. 169-174. Moscow: Institute for System Studies. (In Russian)

55. Terwilliger, P. (1983). Distance-regular graphs and (s,c,a,k)-graphs. J Combin. Theory (B). 34, 151-164.

56. Tutte, W.T. (1947). A family of cubical graphs. Proc. Cambridge Phil. Soc., 43, 459-474.

57. Weiss, R. (1979). Elations of graphs. Acta Math. Hungaricae. 34, 101-103.

58. Weiss, R. (1981). s-Transitive graphs. In Algebraic Methods in Graph Theory, vol. 2. pp. 827-847. Amsterdam: North-Holland.

59. Weiss, R. (1981). The nonexistence of 8-transitive graphs. Combinatorica. 1, 309-311.

60. Weiss, R. (1985). On distance-transitive graphs. Bull. London Math. Soc., 17, 253-256.

61. Weiss, R. (1985). Distance-transitive graphs and generalized polygons. Arch. Math., 45, 186-192.

62. Wong, W.J. (1967). Determination of a class of primitive permutation groups. Math. Z., 99, 235-246.

63. Yanushka, A. (1976). Generalized hexagons of order (t,t). Israil. J. Math., 23, 309-324.

64. Yokoyama, K. (1986). On distance-transitive graphs in which the stabilizer of a point contains an alternating group. Europ. J. Comb. Submitted.

65. Zaytchenko, V.A., Klin, M.H. & Faradžev, I.A. (1980). On some questions concerning the representation of permutation groups in the memory of computer. In Computations in Algebra, Number Theory and Combinatorics. pp. 21-32. Kiev. (In Russian)

66. Zaytchenko, V.A. (1981). Algorithmic approach to the synthesis of combinatorial objects and computation in permutation groups using the method of invariant relations. Ph. D. Thesis. Moscow, MFTI. (In Russian)

67. Zaytchenko, V.A., Ivanov, A.A. & Klin M.H. (1982) Construction and inverstigation of some new automorphic graphs. In Methods and programms of solu-

tion of optimization problems on graphs and
networks. pp. 48-50. Novosibirsk. (In Russian)

LATIN SQUARE DETERMINANTS

K.W. Johnson
The Pennsylvania State University, Ogontz Campus
Abington, PA 19001 U.S.A.

1 INTRODUCTION

Given any latin square, for example

$$
\begin{array}{ccc}
1 & 2 & 3 \\
2 & 3 & 1 \\
3 & 1 & 2
\end{array},
$$

the <u>latin square determinant</u> is defined by replacing each i in the square by a variable x_i and forming the corresponding determinant

$$
\begin{vmatrix}
x_1 & x_2 & x_3 \\
x_2 & x_3 & x_1 \\
x_3 & x_1 & x_2
\end{vmatrix}.
$$

Thus the determinant is a homogeneous polynomial of degree n in n variables where n is the size of the square.

When the latin square is the unbordered multiplication table of a group the problem of finding the irreducible factors of the determinant was suggested by Dedekind to Frobenius and led to the discovery by the latter of group character theory, on the way to the description of a precise algorithm which produces the factorisation of any group determinant given the group table and the character table (see Frobenius (1896, a, b), Hawkins (1971, 1974) and Section 2). It is the purpose of this paper to discuss the problem when the group condition is dropped. Since up to sign the determinant is essentially invariant under isotopy (i.e. row permutations, column permutations and the renaming of elements) the problem may be restricted to the case where the square is the unbordered multiplication table of a loop, or a related square which is described below. Thus the theory is given in terms of loop determinants.

The theory of quasigroup characters as expounded in Johnson & Smith (1984, 1986, 1987) is used to obtain information on the factors, but as may be expected the results obtained are not as complete as in the group case.

In Section 2 a brief account of Frobenius' results is given. The equation (2.2) plays an important role in the group case, so in Section 3 it is shown that associativity is a necessary and sufficient condition for (2.2) to hold. There remains the interesting question of whether associativity is necessary for (2.3) to hold. In Section 4 the reduced loop determinant is defined and it is shown that the decomposition of this into linear factors follows from the standard theory of association schemes. A summary of the results which hold in the general case is given in Section 5, in particular it may be mentioned that the number of irreducible factors of a loop determinant is greater than or equal to the number of conjugacy classes of the loop. In Section 6 there are given some examples calculated on a computer and these provide some counterexamples to show that several of Frobenius' results do not carry over. The work has raised some questions which seem to be of interest and these are presented in Section 7 together with some remarks and conclusions.

2 THE GROUP CASE

The group determinant of a finite group G is defined to be the determinant Θ of the matrix X whose rows and columns are indexed by the elements of G and whose $(g,h)^{th}$ entry is $x_{gh^{-1}}$. This matrix is obtained from that arising directly from the group multiplication table by permuting the columns so that x_1 appears on the diagonal and corresponds to a quasigroup isotopic to the group. Up to sign the two determinants obviously coincide. However the matrix X as defined above has the following important property: let the matrices Y and Z be defined analogously, i.e. $Y=\{y_{gh^{-1}}\}$ and $Z=\{z_{gh^{-1}}\}$, with the further relation

$$z_a = \sum_{bc=a} x_b y_c \tag{2.1}.$$

Then

$$X.Y = Z \tag{2.2},$$

and in particular

$$detX.detY = detZ \qquad (2.3).$$

It is property (2.3) which is used extensively by Frobenius in his proofs of the basic results on group characters and the group determinant.

The following is a summary of Frobenius' main results.

(1) The number of irreducible factors of Θ is equal to the number of conjugacy classes of G.

(2) Each irreducible factor Φ of Θ gives rise to an irreducible character X_Φ of G in the following explicit manner: $X_\Phi(g)$ is f times the coefficient of $x_1^{f-1}x_g$ in Φ, where f is the degree of Φ, and g is an arbitrary element of G.

(3) Given an irreducible character X of G, the corresponding irreducible factor Φ_X is produced by

$$f!\Phi_X = \Sigma X(g_1,g_2,\ldots,g_f)x_{g_1} x_{g_2} \ldots x_{g_f} \qquad (2.4),$$

where $X(g_1,g_2,\ldots,g_f)$ is defined recursively by

$$X(g_1,g_2,\ldots,g_f) = X(g_1)X(g_2,g_3,\ldots,g_f) - X(g_1 g_2,g_3,\ldots,g_f)$$
$$-\ldots -X(g_2,g_3,\ldots,g_1 g_f) \qquad (2.5).$$

where $f=X(e)$ and $(g_1,g_2,\ldots,g_f)$ runs over all f-tuples of elements of G.

(4) If Φ is an irreducible factor of Θ then the highest power of Φ which divides Θ is f, the degree of Φ.

(5) Define the <u>reduced determinant</u> Θ_R of G by making the identification $x_g = x_h$ in Θ whenever g and h lie in the same conjugacy class of G. Then Θ_R factorises into linear factors and if Ψ is such a factor with

$\Psi = \underset{\lambda \epsilon G}{\Sigma} \lambda(g)x_g$ the corresponding character X is obtained as

$X(g) = [\lambda(g)/n(g)].f_X$ where $n(g) = |C(g)|$, the number of conjugates of g, and f_X is defined by $f_X^2 . \underset{g \epsilon G}{\Sigma} (\lambda(g)^2/n(g)) = |G|$.

3 THE EQUATION X.Y=Z IN THE GENERAL CASE

Let Q be a finite loop. Let X be the matrix with rows and columns indexed by the elements of Q with the $(g,h)^{th}$ entry $x_{gp(h)}$ where $\rho(h)$ is the right inverse of h and let Y and Z be defined similarly. Let the <u>loop determinant</u> Θ be defined by $\Theta = \det(X)$. Suppose that (2.1) is satisfied.

<u>Proposition 3.1.</u> The equation X.Y=Z holds if and only if Q is associative.

<u>Proof.</u> (a) (Frobenius' proof for groups). Let Q be a group. Then the $(g,h)^{th}$ entry of X.Y is

$$\sum_{k \in G} x_{gk^{-1}} y_{kh^{-1}} = \sum_{bc=gh^{-1}} x_b y_c = z_{gh^{-1}},$$

and thus X.Y=Z.

(b) Suppose X.Y=Z holds. Then as above the $(g,h)^{th}$ entry of X.Y is $\sum_{k \in Q} x_{g\rho(k)} y_{k\rho(h)}$. The product $x_{g\rho(k)} y_{k\rho(h)}$ occurs in the expression for $z_{g\rho(h)} = \sum_{bc=g\rho(h)} x_b y_c$ only if

$$[g\rho(h)].[k\rho(h)] = g\rho(h) \tag{3.1}.$$

and thus (2.2) holds only if (3.1) holds for all g,h,k in Q. Now if g=h=e, (3.1) becomes $\rho(k).k=e$, i.e. the left and right inverses coincide in Q and $\rho(k)$ may be written as k^{-1}. Equation (2.1) then becomes

$$(gk^{-1})(kh^{-1}) = gh^{-1} \tag{3.2}.$$

If h=e in (3.2) we obtain

$$(gk^{-1})k = g \tag{3.3}.$$

Now let $gh^{-1}=c$. By (3.3), $g=(gh^{-1})h = ch$. Thus insertion in (3.2) produces $c(hk^{-1})=(ch)k^{-1}$ and since c,k,h are arbitrary elements of Q, Q must be associative.

I do not know whether (2.3) implies the associativity of Q.

4 THE REDUCED LOOP DETERMINANT

As in Section 3 let Q be a finite loop. A conjugacy class of Q is defined to be a subset of QxQ (see Johnson & Smith (1984)) and I define the reduced matrix X_R by identifying the $(g,h)^{th}$ and $(k,m)^{th}$ entries of X whenever (g,h) and (k,m) lie in the same conjugacy class of Q. It is straightforward to see that this definition coincides with that in Section 2 for groups. Let Θ_R be $\det(X_R)$.

Now suppose that Q has basic characters $X_1, X_2, \ldots, X_r$ as defined in Johnson & Smith (1984).

<u>Proposition 4.1</u>. Θ_R factorises into linear factors. There is a bijection between these factors and $X_1, X_2, \ldots, X_r$. The multiplicity with which the linear factor μ_i appears in Θ_R is equal to the square of the degree of the corresponding X_i.

<u>Proof</u>. Note that if $g\rho(h) = k\rho(m)$ the elements (g,h) and (k,m) must lie in the same conjugacy class. For $(g,h)\rho(h) = (g\rho(h),e) = (k\rho(m),e) = (k,m)\rho(m)$ and hence (g,h) and (k,m) lie in the same orbit under the diagonal action of G. The matrix X_R can be decomposed as $X_R = \sum_{k=1}^{r} x_k A_k$ where the A_k are the basis matrices for the Bose-Mesner algebra which arises from Q (see Johnson & Smith (1984)) and x_k is the variable which corresponds to the kth conjugacy class of Q. Now the A_k can be simultaneously diagonalised so that A_k is similar to diag$(\lambda_{k_1}, \lambda_{k_2}, \ldots, \lambda_{k_2}, \ldots, \lambda_{k_r}, \ldots, \lambda_{k_r})$ where λ_{k_i} appears f_i times and where f_i is the square of the value of X_i on the trivial class, and hence X_R is similar to a diagonal matrix and factors into linear factors of which a typical one is $\sum_{k=1}^{r} \lambda_{k_i} x_k$, corresponding to X_i and appearing with multiplicity f_i.

5 THE RESULTS IN THE GENERAL CASE

Here again Q denotes a finite loop and Θ its loop determinant.

<u>Theorem 5.1</u>. Let Φ be an irreducible factor of Θ. Then if Φ_R denotes the polynomial obtained by identifying x_g and x_h whenever (e,g) and (e,h) lie in the same conjugacy class of Q, Φ_R is a power of a linear factor of Θ_R.

<u>Corollary 5.2</u>. The number of irreducible factors of Θ is greater than or equal to the number of conjugacy classes of Q.

<u>Corollary 5.3</u>. Any linear character of Q gives rise to a linear factor of Θ.

<u>Corollary 5.4</u>. Each homomorphism of Q onto an abelian group gives rise to a linear factor of Θ.

The details of the proofs will appear elsewhere. Theorem 5.1 depends on a result of Frobenius on commuting matrices with polynomial entries. Corollary 5.2 is an immediate consequence, and Corollary 5.3 follows fairly easily. Corollary 5.4 follows from 5.3 and results on pullbacks of characters along homomorphisms which appear in Johnson & Smith (1987).

6 <u>EXAMPLES</u>

It is possible to compute the character tables and the loop determinants for all isotopy classes of loops of orders 5 and 6. For higher orders the number of such classes becomes forbidding--there are more than 1/4 million such classes of loops of order 8, so the calculations were restricted to a few interesting loops of order 8. The calculations were carried out using MACSYMA, the symbolic manipulation package, on the VAX at Villanova with the help of R. Beck. It was difficult to calculate the factors of the loop determinants of order 8 by the straightforward algorithms, and it was necessary to deduce factors by substituting primes for variables, to guess factors and then to confirm by division. Note that the character table information gives the linear factors, and the degrees of other factors which may or may not be irreducible. A much better method is needed for loops of higher orders (for instance it would be interesting to calculate the factors of the loop determinant of the smallest non-associative Moufang loop which has order 12) (added in proof: this has now been accomplished and the work will be reported separately). Loops of order 4 or less are associative.

In the following a factor is given explicitly if it can be written in a form which is reasonably concise. Since the computer algorithm factors over $\mathbb{Q}$ irreducible means irreducible over $\mathbb{Q}$. The determinant of a loop of order n always has the factor $\sum_{i=1}^{n} x_i$ and this is omitted in each case. The character table of a loop is trivial if there are two conjugacy classes, see Johnson & Smith (1986) section 5.

(1) The loops of order 5. There is one isotopy class, and the character table is trivial. There is one non-trivial factor of the loop de-

terminant (of degree 4).

(2) The loops of order 6. The character tables of these loops are

$$
\begin{array}{ccc}
1 & 1 & 1 \\
1 & 1 & -1 \\
2 & -1 & 0
\end{array}
\qquad \text{(the table of } S_3\text{)}
$$

$$
\begin{array}{cccc}
1 & 1 & 1 & 1 \\
1 & 1 & \omega & \omega^2 \\
1 & 1 & \omega^2 & \omega \\
\sqrt{3} & -\sqrt{3} & 0 & 0
\end{array}
\qquad \text{(the table of 4.1.1)} \qquad (\omega = e^{2\pi i/3})
$$

and the trivial table.

The loops of order 6 are listed in Denes & Keedwell (1964), pp. 130-137 and I use the notation there. They are grouped according to their character tables.

(a) Loops with the character tables of S_3.

S_3: the factors are $(u-v)(u_1 u_2 - v_1 v_2)^2$, where $u = x_1 + x_2 + x_3$, $v = x_4 + x_5 + x_6$, $u_1 = x_1 + \omega x_2 + \omega^2 x_3$, $u_2 = x_1 + \omega^2 x_2 + \omega x_3$, $v_1 = x_4 + \omega x_5 + \omega^2 x_6$ and $v_2 = x_4 + \omega^2 x_5 + \omega x_6$.

3.1.1: the factors are $u-v$ and an irreducible factor.

9.1.1: the factors are $u-v$ and an irreducible factor.

9.2.1: the factors are $(u-v)(u_1 u_2 - v_1 v_2)(u_1 u_2 + v_1 v_2)$.

9.3.1: same as 9.2.1.

(b) The loop 4.1.1: This has the unique character table given above, up to isotopy. Its determinant has the factors

$(x_1 + x_2 + \omega(x_3 + x_4) + \omega^2(x_5 + x_6))$, $(x_1 + x_2 + \omega^2(x_3 + x_4) + \omega(x_5 + x_6))$ and

$\alpha^3 + \beta^3 + \gamma^3 + \alpha\beta\gamma$ where $\alpha = x_1 - x_2$, $\beta = x_4 - x_3$ and $\gamma = x_5 - x_6$.

(c) Loops with the trivial character table. If a loop is not given its determinant has an irreducible factor of degree 5.

5.1.1. the factors are $(t_1 w_1 - t_2 w_2)$ and an irreducible factor of degree 3, where $t_1 = x_1 + \omega x_3 + \omega^2 x_4$, $t_2 = x_2 + \omega x_5 + \omega^2 x_6$, $w_1 = x_1 + \omega^2 x_3 + \omega x_4$ and $w_2 = x_2 + \omega^2 x_5 + \omega x_6$.

10.1.1. the factors are $(r_1 s_1 + r_2 s_2)$ and an irreducible factor of degree 3, where $r_1 = x_1 + \omega x_5 + \omega^2 x_6$, $r_2 = x_2 + \omega x_3 + \omega^2 x_4$, $s_1 = x_1 + \omega^2 x_5 + \omega x_6$ and $s_2 = x_2 + \omega^2 x_3 + \omega x_4$.

(3) The loops of order 8. There are two isotopy classes of Bol loops
of order 8 and these have the same character table as that of D_8, the
dihedral group of order 8, and Q_8, the quaternion group of that order.
In addition there is an A-loop of order 8, i.e. a loop in which all
inner maps are automorphisms, with the same table. The following are
the latin squares corresponding to the three isotopism classes of loops:

<pre>
            (Bol 1)   (Bol 2)
    1 2 3 4 5 6 7 8   1 2 3 4 5 6 7 8
    2 1 4 3 6 5 8 7   2 1 4 3 6 5 8 7
    3 4 1 2 8 7 5 6   3 4 2 1 7 8 6 5
    4 3 2 1 7 8 6 5   4 3 1 2 8 7 5 6
    5 6 7 8 1 2 3 4   5 6 7 8 2 1 3 4
    6 5 8 7 2 1 4 3   6 5 8 7 1 2 4 3
    7 8 5 6 4 3 1 2   7 8 6 5 4 3 2 1
    8 7 6 5 3 4 2 1   8 7 5 6 3 4 1 2

                    (A-loop)
            1 2 3 4 5 6 7 8
            2 1 4 3 6 5 8 7
            3 4 2 1 8 7 5 6
            4 3 1 2 7 8 6 5
            5 6 7 8 1 2 4 3
            6 5 8 7 2 1 3 4
            7 8 6 5 3 4 1 2
            8 7 5 6 4 3 2 1
</pre>

It follows directly from the character table that all five determinants
have the following three non-trivial linear factors: $(u+v-t-w)$, $(u-v+t-w)$
and $(u-v-t+w)$, where $u=x_1+x_2$, $v=x_3+x_4$, $t=x_5+x_6$, and $w=x_7+x_8$.

Let $\alpha=x_1-x_2$, $\beta=x_3-x_4$, $\gamma=x_5-x_6$, $\delta=x_7-x_8$. Then the remaining factors are as follows:

D_8: $(\alpha^2+\beta^2-\gamma^2-\delta^2)^2$,

Q_8: $(\alpha^2+\beta^2+\gamma^2+\delta^2)^2$,

Bol 1: $(\alpha^2-\beta^2-\gamma^2-\delta^2)(\alpha^2+\beta^2-\gamma^2-\delta^2)$,

Bol 2: $(\alpha^2+\beta^2+\gamma^2+\delta^2)(\alpha^2+\beta^2-\gamma^2-\delta^2)$,

A-loop: an irreducible factor of degree 4.

7 CONCLUDING REMARKS.

In the case where G is a group the reduced determinant is essentially equivalent to the character table of G. The full determinant is a stronger invariant (see the results in Section 6 for D_8 and Q_8), and the following appears to be open.

<u>Problem 1</u>. If groups G_1 and G_2 have the same group determinant are they necessarily isomorphic?

There is an obvious generalisation. The loop determinant is invariant under isotopy and reversal of multiplication $(x*y = y.x)$.

<u>Problem 2</u>. If loops Q_1 and Q_2 have the same loop determinant are they necessarily isotopic or trivially related in some other sense?

From the examples in Section 6 it is clear that there can be a strictly greater number of factors of the determinant than the number of conjugacy classes, even in the case when the loop character table is trivial. It is easily shown that even for characters of degree 2 Frobenius' algorithm for producing the corresponding irreducible factor does not go through (e.g. for the loop Bol 1 described in Section 6), and it appears that any algorithm to produce factors of loop determinants in the same manner would be difficult to obtain. A more tractible problem may be the following.

<u>Problem 3</u>. If the loop determinant of a loop Q has a linear factor, does this arise necessarily from a linear character?

<u>References</u>.

Denes, J. & Keedwell, A.D. (1964). Latin Squares And Their Applications, Academic Press.

Frobenius, G. (1896). Uber Gruppencharaktere, S'ber Akad. d. Wiss. Berlin, 985-1021.

Frobenius, G. (1896). Uber die Primfaktoren der Gruppendeterminante, S'ber. Akad. d. Wiss. Berlin, 1343-1382.

Hawkins, T. (1971). The Origins of the Theory of Group Characters, Arch. Hist. Exact Sci. 7, 142-170.

Hawkins, T. (1974). New Light on Frobenius' Creation of the Theory of Group Characters, Arch. Hist. Exact Sci. 12, 217-243.

Johnson, K.W. & Smith, J.D.H. (1984). Characters of Finite Quasigroups, Europ. J. Combinatorics 5, 43-50.

Johnson, K.W. & Smith, J.D.H. (1986). Characters of Finite Quasigroups II: Induced Characters, Europ. J. Combinatorics 7 (2) 131-138.

Johnson, K.W. & Smith, J.D.H. (1987). Characters of Finite Quasigroups III: Quotients and Fusion, submitted for publication.

A COMPUTER SEARCH FOR A PROJECTIVE PLANE OF ORDER 10

C.W.H. Lam
Department of Computer Science, Concordia University,
Montreal, Quebec, H3G 1M8

L.H. Thiel
Department of Computer Science, Concordia University,
Montreal, Quebec, H3G 1M8

S. Swiercz
Department of Computer Science, Concordia University,
Montreal, Quebec, H3G 1M8

0 *ACKNOWLEDGEMENT*

This work was supported by the Natural Sciences and Engineering Research Council of Canada under Grants A9373, 0011 and by the Fonds pour la Formation de Chercheurs et l'Aide á la Recherche under Grant EQ2369.

1 *INTRODUCTION*

A finite projective plane of order n is a collection of n^2+n+1 lines and n^2+n+1 points such that

(1.1)   every line contains $n+1$ points,

(1.2)   every point is on $n+1$ lines,

(1.3)   any two distinct lines intersect at exactly one point, and

(1.4)   any two distinct points lie on exactly one line.

For example, a projective plane of order 2 is shown in Fig. 1. It has 7 points and 7 lines. The points are numbered from 1 to 7. The 7 lines are $L_1=$ {1,2,4}, $L_2=$ {2,3,5}, $L_3=$ {3,4,6}, $L_4=$ {4,5,7}, $L_5=$ {1,5,6}, $L_6=$ {2,6,7} and $L_7=$ {1,3,7}. In Fig. 1 all the lines except L_6 are drawn as straight lines. One can easily show that this projective plane of order 2 is unique up to the relabelling of points and lines.

Another way to represent a projective plane is to use an *incidence matrix* A of size n^2+n+1 by n^2+n+1. The columns represent the points and the rows represent the lines. The entry A_{ij} is 1 if point j is incident on line L_i, otherwise it is 0. Figure 2 gives the incidence matrix corresponding to the projective plane of order 2 shown in Fig. 1.

It is clear that A is a (0,1)-matrix. The conditions (1.1) to (1.4) are translated into the following conditions:

(1.5) A has constant row sum $n+1$,

(1.6) A has constant column sum $n+1$,

(1.7)   the inner product of any two distinct rows of A is 1, and

(1.8)   the inner product of any two distinct columns of A is 1.

It is known that if n is a prime or a prime power, then a plane of order n exists. The construction depends essentially on the existence of a finite field of that order. The first value of n which is not a prime power is 6. Tarry (1900) proved that a projective plane of order 6 does not exist. The next value is $n = 10$. The question whether a projective plane of order 10 exists or not has been open for over 200 years. This paper reports the results of a computer search for such a plane. It also discusses some of the programming techniques used to reduce the size of the search. All indications are that a computer solution to the problem is definitely feasible. It is quite possible that in three years time, an answer to the existence question of a projective plane of order 10 will be known.

Fig. 1   Projective plane of order 2

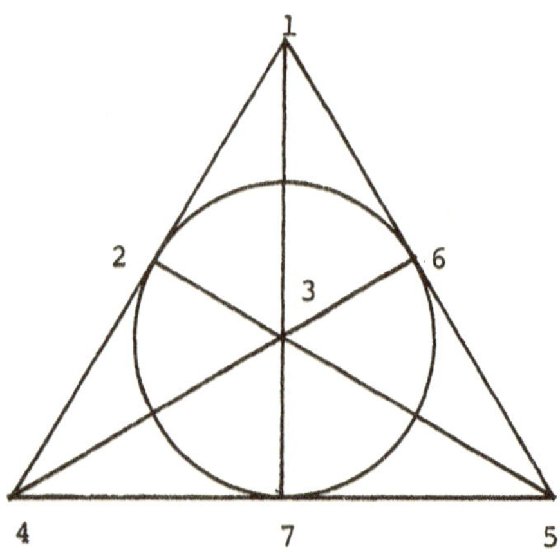

Fig. 2   Incidence matrix for a projective plane of order 2

	1	2	3	4	5	6	7
L_1	1	1	0	1	0	0	0
L_2	0	1	1	0	1	0	0
L_3	0	0	1	1	0	1	0
L_4	0	0	0	1	1	0	1
L_5	1	0	0	0	1	1	0
L_6	0	1	0	0	0	1	1
L_7	1	0	1	0	0	0	1

2 AN ERROR CORRECTING CODE

MacWilliams, Sloane and Thompson (1973) introduced the idea of studying the binary error correcting code associated with a projective plane of order 10. Let A be the incidence matrix of such a plane. Let S be the vector space generated by the rows of A over GF(2). A vector in S is called a *codeword*. The *weight* of a codeword is the number of 1's in the codeword. Let w_i be the number of codewords of weight i. We define the *weight enumerator* of S to be

$$\sum_{i=0}^{111} w_i\, x^i.$$

In MacWilliams et al. (1973), it was noted that the weight enumerator of S is uniquely determined once w_{12}, w_{15} and w_{16} are known. They showed, using a computer, that $w_{15} = 0$. They also proved the following result.

Theorem 1. [MacWilliams, Sloane, Thompson] If $i \equiv 1$ or $2 \pmod 4$, then
$$w_i = 0.$$

They also proved that codewords of weight 12 are exactly the ovals of the plane. Here, an *oval* in a projective plane of order 10 is a set of 12 points, no 3 of which are collinear.

2.1 The oval configuration

Since no three points of an oval are collinear, a line intersects an oval at most twice. A line cannot intersect an oval at exactly 1 point; otherwise, the oval with

Fig. 3. Oval Configuration

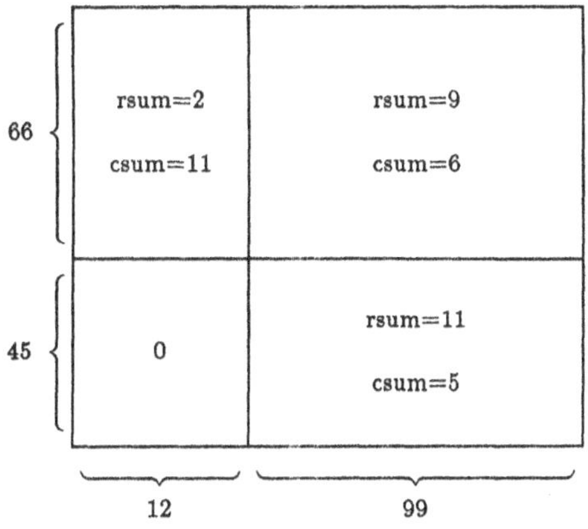

66
rsum=2
csum=11
rsum=9
csum=6
45
0
rsum=11
csum=5
12
99

this line will form a codeword of weight 21 which is impossible by Theorem 1. Thus a
line intersects the oval in either 0 or 2 points. Therefore the $\binom{12}{2} = 66$ pairs of points
in the oval each define a unique line. A point on the oval is on 11 of these 66 lines.
A point outside the oval is on 6 of these lines.

The above discussion implies that if there exists an oval in a projective
plane of order 10, then its incidence matrix can be permuted, by independent row and
column permutations, to the form shown in Fig. 3. Here, the first 12 columns represent
the 12 points of the oval. The first 66 rows represent the 66 lines each of which
contains 2 points of the oval. The notations rsum and csum denote the row sum and
column sum, which are the number of ones in a (partial) row or column respectively.
The values of the rsum and csum are all derived from the previous discussion about the
incidence of lines and points. For example, csum is 6 on the top right region because
a point outside the oval is on 6 of the 66 lines.

Basically, our computer search for ovals in a projective plane of order 10
uses exhaustive backtrack search to try to construct the top 66 rows of an incidence
matrix of the form in Fig. 3, satisfying the conditions (1.5) to (1.8). *If one cannot
find, after an exhaustive search, an incidence matrix of the form in Fig. 3, then there is
no oval in a projective plane of order 10.*

In 1983, we completed such a search without finding an incidence matrix
of the required form (Lam et al., 1983). The program used 183 days of CPU time on
a VAX-11/780. Thus, we know that $w_{12} = 0$.

2.2   *The 16-point configurations*

Let v be a set of 16 points which form a codeword of weight 16. One
can show (Carter, 1974, and Hall, 1980) that 8 of the 111 lines intersect v in 4 points.

Fig. 4   Carter's Case I

```
1 1 1 1   0 0 0 0   0 0 0 0   0 0 0 0
0 0 0 0   1 1 1 1   0 0 0 0   0 0 0 0
0 0 0 0   0 0 0 0   1 1 1 1   0 0 0 0
0 0 0 0   0 0 0 0   0 0 0 0   1 1 1 1
1 0 0 0   1 0 0 0   1 0 0 0   1 0 0 0
0 1 0 0   0 1 0 0   0 1 0 0   0 1 0 0
0 0 1 0   0 0 1 0   0 0 1 0   0 0 1 0
0 0 0 1   0 0 0 1   0 0 0 1   0 0 0 1
```

Fig. 5   Carter's Case VI

```
1 0 0 0   1 1 1 0   0 0 0 0   0 0 0 0
0 0 0 0   0 0 0 0   1 0 1 0   0 1 0 1
0 0 0 0   0 0 0 1   0 0 0 1   0 0 1 1
1 1 1 1   0 0 0 0   0 0 0 0   0 0 0 0
0 0 0 0   0 0 1 0   0 1 0 0   1 0 1 0
0 1 0 0   1 0 0 1   1 0 0 0   0 0 0 0
0 0 1 0   0 1 0 0   0 1 1 0   0 0 0 0
0 0 0 1   0 0 0 0   0 0 0 1   1 1 0 0
```

These lines are called *heavy* lines.   72 of the remaining lines intersect v at 2 points and the other 31 lines do not intersect v.

Carter (1974) showed that there are 6 possible incidence structures involving the 8 heavy lines and the 16 points of v.   He finished a computer search for 4 of the cases.   Figures 4 and 5 give the two cases, Case I and Case VI, that remain to be considered.

These two cases can be divided into 6 subcases.   Carter has finished one of the subcases arising from his case VI.   In 1985, we finished the remaining 5 subcases and did not find any weight 16 codeword.   The program used 78 CPU days on a VAX-11/780.   The results were reported in Lam et al. (1986).   Thus, we also know that $w_{16} = 0$.

Since w_{12}, w_{15} and w_{16} are all known, one can compute the weight enumerator.   In particular, if a projective plane of order 10 exists, then it must contain 24,675 codewords of weight 19.   Hence, the question of the existence of this projective plane can be settled by searching for 19-point configurations.

2.3   *The 19-point configurations*

Let v be a set of 19 points which form a codeword of weight 19. Theorem 1 implies that every line must intersect v at an odd number of points.   An intersection of 7, 9 or 11 points implies the existence of a weight 16, 12 or 8 codeword; all of which are impossible.   Thus, the possible intersection numbers are 1, 3 or 5. Hall (1980) showed that there are 6 *heavy* lines each containing 5 points of v, 37 *triple* lines each containing 3 points of v, and 68 *single* lines each containing 1 point of v.

There are 66 different incidence structures of the 6 heavy lines with the 19 points of v.   These are our starting cases.   In Lam et al. (1985), 21 of these cases are eliminated by theoretical arguments.   A further 37 cases were eliminated by computer runs between May 1985 and February 1987.   Since we are starting our computer search with the easy cases first, the remaining 8 cases are getting more and more difficult.   In particular, we estimated that the most difficult case, case 6, would require over 5 years of CPU time on a VAX-11/780.   We have developed a version of the program to be run on a CRAY-1.   As of April, 1987, we have finished three of these 8 cases.   We estimate that, together, the remaining 5 cases will require about 3 months of CPU time.

In the next section, we shall discuss some of the computing techniques used to speed up the computer search so that it becomes feasible.

3   COMPUTATIONAL ASPECTS

3.1   Backtrack search

First of all, let us define the search problem in a more general context.

Search problem:

Given a collection of sets of candidates C_1, C_2, C_3, ..., C_n and a boolean compatibility function $f(x,y)$ defined for all $x \in C_i$ and $y \in C_j$, find a transversal T of size n with exactly one element form each C_i such that for all pairs x, $y \in T$, $x \neq y$, $f(x,y)$ is true.

For example, if we take C_i to be the set of all candidates for column i of the incidence matrix of a projective plane, then $f(x,y)$ can be defined as

$$f(x,y) \;=\; \begin{cases} \text{true} & \text{if } \langle x,y \rangle \;=\; 1 \\ \text{false} & \text{otherwise,} \end{cases}$$

where $\langle x,y \rangle$ denotes the inner product of x and y. A transversal is then a complete incidence matrix.

On the other hand, we may be only interested in a partial incidence matrix such as the first 66 rows of Fig. 3. If C_i is the set of all candidates for a partial column i of the incidence matrix, then $f(x,y)$ can be defined as

$$f(x,y) \;=\; \begin{cases} \text{true} & \text{if } \langle x,y \rangle \;\leq\; 1 \\ \text{false} & \text{otherwise.} \end{cases}$$

A transversal is now only a partial incidence matrix.

A general backtrack search program can now be stated as follows:

```
For each x_1 in C_1 do
begin
        For each x_2 in C_2 compatible to x_1 do
        begin

                .
                .

                For each x_i in C_i compatible to x_1,x_2...,x_{i-1} do

                .
                .
                .

        end;
end;
```

Such a backtrack search program can best be written recursively. The next subsection discusses how one estimates the amount of computing time required by a backtrack search.

3.2 *Estimation*

Knuth (1975) introduced a Monte Carlo method to estimate the amount of computing time required by a backtrack program. It is based on estimating the number of nodes at each level of the search tree corresponding to the backtrack program. In our context, a node at level k is a partial transversal $T = \{x_1,...,x_k\}$ whose elements are mutually compatible. His idea is to run a number of experiments, each consisting of performing the backtrack search with a randomly chosen candidate at each level of the search. Suppose we have a partial transversal $\{x_1,...,x_k\}$. We choose x_{k+1} at random among the set

$$S_{k+1} = \{y \in C_{k+1}, \ y \text{ compatible to } x_1,x_2...,x_k\}.$$

We define $n_k = |S_k|$. An estimate for the number of nodes at level k, α_k, is the average value of $n_1 n_2...n_k$. An estimate for the total number of nodes, E, in the search tree is

$$E = \sum_{i=1}^{n} \alpha_i \ .$$

The estimated CPU time required is then $E \times$ (CPU time required to process one node).

These estimated values of the node counts can best be presented by plotting the logarithm of α_i (the number of digits in α_i) as a function of i. We call this plot the *profile* of the search. A typical profile is shown in Fig. 6. The value of i for which $\log \alpha_i$ is maximum is called the *bulge* of the search. A backtrack search program spends most of its time processing nodes near the bulge.

We have developed a prototype program, called NPL, which search for the incidence matrix of a projective plane of order 10. It is slow, at least 100 times slower than the final versions of the program. However, it is easily adaptable to new configurations. We believe that the NPL program is one of the major reasons which makes our project feasible. Besides telling us whether the problem can be solved with our present day computers, we also use it to quickly explore ideas of "improvement" to the search. Ideas that turn out to increase the estimated CPU time are discarded. The program also gives us the profile of the search, which we use to design our optimized final program. We also use the program to check the correctness of our final program. Selected small subcases are run on both programs and the intermediate results are compared.

3.3   *Optimization*

In this subsection, we describe some of the techniques used to reduce the required CPU time. These techniques can be divided into two broad classes:

a.   those that reduce the size of the search tree, and

b.   those that reduce the time required to process each node.

The first idea to reduce the size of a search tree is to use isomorph rejection. For example, let us consider Carter's Case I shown in Fig. 4. One of the subcases is the "two distinguished point" case where there are two points each of which is incident on 4 of the 8 heavy lines. The order of the automorphism group of the 8 × 16 submatrix in Fig. 4 is 1,152. The subgroup fixing the first line is of order 144. We know the incidence structure of 5 of the 11 points on this line (4 in the codeword and 1 distinguished point). We call the remaining 6 points (columns) on this line B_1, for block 1. The subgroup of size 144 reduces the number of B_1's from 49,472 to 469 non-isomorphic cases. Furthermore, the original group is transitive on the 8 heavy lines. Parts of this symmetry can be recovered by a method of pregeneration and tags. Please see Lam et al. (1986) for more detail.

The second idea to reduce the size of a search tree is to use a better ordering of the candidate sets C_i's. Again, we use the two distinguished point subcase of Carter's Case I as an example. In the area of the partial columns where we were working, it was found that columns (points) outside the codeword but incident on one of the heavy lines can be divided into two classes, those incident on one of the first 4 lines and those incident on one of the second 4 lines. A column candidate is more likely to be compatible to another column candidate from the same class than those

Fig. 6   Typical profile of a search

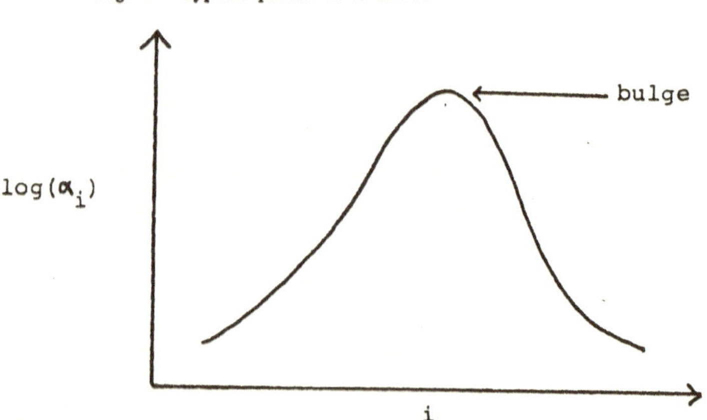

from the other class. Thus in our search, after completing B_1, we switch to work on columns incident on row 7, then those in row 8 and then back to those in row 4.

The size of the search tree can also be reduced by changing the set of candidates. In fact, in most of our programs, the candidates are partial columns. For example, in our program to search for ovals, columns are of size 66. In other words, we are only concerned with the incidence structure of points and the 66 lines containing points of the oval. In general, we try to concentrate on regions of the incidence matrix with a high density of ones, because it tends to lead to incompatibility and hence a slimmer search tree.

We can also reduce the amount of execution time by processing each node faster. One possibility is to use a faster computer. We are making arrangements to use a CRAY-1 because we realized from our estimation that our programs will take too long on our VAX-11/780.

Fortunately, the backtrack search process can be implemented with bit-vector operations which makes the CRAY an ideal machine for the problem.

The reason that bit-vectors are used is because we want to do as little as possible near the bulge. The method uses a *compatibility matrix* which is a term introduced by Carter (1974). We will explain the method with an example.

The 4-queens' problem is to find the placement of 4 queens on a 4×4 chessboard such that no two queens attack one another. We know that there must be one queen in each column. In each column, the queen can be in any one of the 4 rows. Thus, each of the 4 candidate sets (one for each column) has 4 elements. We

Fig. 7.   Compatibility matrix for the 4-queens' problem

$$
D = \begin{bmatrix}
0 & 0 & 0 & 0 & 0 & 0 & 1 & 1 & 0 & 1 & 0 & 1 & 0 & 1 & 1 & 0 \\
0 & 0 & 0 & 0 & 0 & 0 & 0 & 1 & 1 & 0 & 1 & 0 & 1 & 0 & 1 & 1 \\
0 & 0 & 0 & 0 & 1 & 0 & 0 & 0 & 0 & 1 & 0 & 1 & 1 & 1 & 0 & 1 \\
0 & 0 & 0 & 0 & 1 & 1 & 0 & 0 & 1 & 0 & 1 & 0 & 0 & 1 & 1 & 0 \\
0 & 0 & 1 & 1 & 0 & 0 & 0 & 0 & 0 & 0 & 1 & 1 & 0 & 1 & 0 & 1 \\
0 & 0 & 0 & 1 & 0 & 0 & 0 & 0 & 0 & 0 & 0 & 1 & 1 & 0 & 1 & 0 \\
1 & 0 & 0 & 0 & 0 & 0 & 0 & 0 & 1 & 0 & 0 & 0 & 0 & 1 & 0 & 1 \\
1 & 1 & 0 & 0 & 0 & 0 & 0 & 0 & 1 & 1 & 0 & 0 & 1 & 0 & 1 & 0 \\
0 & 1 & 0 & 1 & 0 & 0 & 1 & 1 & 0 & 0 & 0 & 0 & 0 & 0 & 1 & 1 \\
1 & 0 & 1 & 0 & 0 & 0 & 0 & 1 & 0 & 0 & 0 & 0 & 0 & 0 & 0 & 1 \\
0 & 1 & 0 & 1 & 1 & 0 & 0 & 0 & 0 & 0 & 0 & 0 & 1 & 0 & 0 & 0 \\
1 & 0 & 1 & 0 & 1 & 1 & 0 & 0 & 0 & 0 & 0 & 0 & 1 & 1 & 0 & 0 \\
0 & 1 & 1 & 0 & 0 & 1 & 0 & 1 & 0 & 0 & 1 & 1 & 0 & 0 & 0 & 0 \\
1 & 0 & 1 & 1 & 1 & 0 & 1 & 0 & 0 & 0 & 0 & 1 & 0 & 0 & 0 & 0 \\
1 & 1 & 0 & 1 & 0 & 1 & 0 & 1 & 1 & 0 & 0 & 0 & 0 & 0 & 0 & 0 \\
0 & 1 & 1 & 0 & 1 & 0 & 1 & 0 & 1 & 1 & 0 & 0 & 0 & 0 & 0 & 0
\end{bmatrix}
\begin{matrix}
1 \\ 2 \\ 3 \\ 4 \\ 5 \\ 6 \\ 7 \\ 8 \\ 9 \\ 10 \\ 11 \\ 12 \\ 13 \\ 14 \\ 15 \\ 16
\end{matrix}
$$

number the squares in the chessboard from 1 to 16, starting from the left and going down the columns. Then square i is compatible with square j if they are not on the same row, column or diagonal. Figure 7 gives the compatibility matrix of the 4-queens' problem. The entries are defined by

$$D_{ij} = \begin{cases} 1 & \text{if } i \neq j \text{ and } f(i,j) = \text{true} \\ 0 & \text{otherwise.} \end{cases}$$

The backtrack search uses a *live choice vector* V which is initially all 1's, or $V = [1111\ 1111\ 1111\ 1111]$. Bit 1 is on. Hence we choose candidate 1 of column 1 and update the live choice vector by "ANDing" it with the first row of D.

The new V is $[0000\ 0011\ 0101\ 0110]$. Bit 7 is on, thus one can place a queen in row 3 of column 2. Suppose we take this choice, the new live choice vector is obtained by "ANDing" it with row 7 of D, which gives $[0000\ 0000\ 0000\ 0100]$. This represents a dead-end because there is no live candidate in column 3.

There are advantages and disadvantages in using the bit-vectoring method. On the positive side, compatibility testing is replaced by fast parallel bit-vector operations. It also avoids the expensive process of generating candidates near the bulge. On the negative side, one has to pregenerate all the candidates as well as the compatibility matrix, which may be very large.

Our strategy is to divide the search into 3 phases as shown in Figure 8. In phases 1 and 3, we use standard backtracking. In phase 2, we use the bit-vectoring method. The boundary line between phases is chosen based on the estimated profile. We have to avoid setting the boundaries too far from the bulge; otherwise the candidate lists are too large. We also have to avoid having the boundaries too close to the bulge; otherwise we do not benefit from the bit-vectoring.

Fig. 8.   Strategy for using the bit-vectoring method

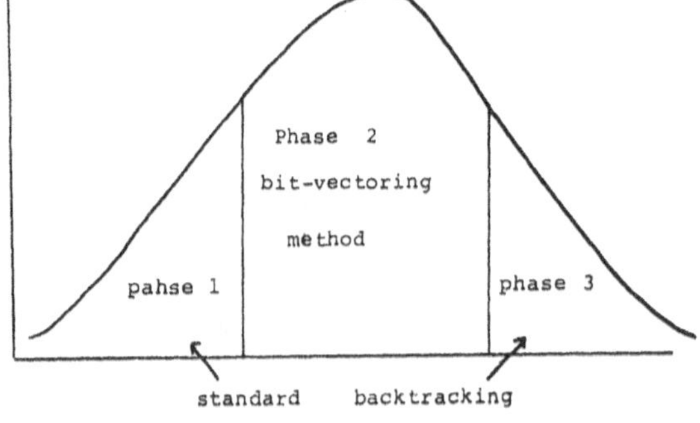

4 CONCLUSION

At this moment, we have most of the programs in place. We estimate that the 8 cases to be run on a CRAY-1 will take about 3 months of CPU-time and about 3 years of elapsed time. If all goes well, the question of the existence of a projective plane should be settled within 3 years.

REFERENCES

Carter, J.L., "On the existence of a projective plane of order ten", Ph.D. Thesis, Univ. of Calif. Berkeley, 1974.

Hall, M., Jr., "Configurations in a plane of order 10", Ann. Discrete Math., 6 (1980) 157-174.

Knuth, D.E., "Estimating the Efficiency of Backtrack Programs", Math. Comp. 29 (1975), 121-136.

Lam, C., Thiel, L., Swiercz S., and McKay, J., "The nonexistence of ovals in a projective plane of order 10", Discrete Math., 45 (1983), 319-321.

Lam, C., Thiel, L., and Swiercz, S., "The Nonexistence of Code Words of Weight 16 in a Projective Plane of Order 10", J. Combin. Theory, See A., 42 (1986), 207-214.

Lam, C., Crossfield, S., and Thiel, L., "Estimates of a computer search for a projective plane of order 10", Congressus Numerantium, 48(1985), 253-263.

MacWilliams, F.J., Sloane, N.J.A., and Thompson, J.G., "On the existence of a projective plane of order 10", J. Combin. Theory, Sec. A., 14(1973), 66-78.

Tarry, G., "Le problème des 36 officiers", C. R. Assoc. France Av. Sci. 29(1900), part 2, 170-203.

MATROIDS, ALGEBRAIC AND NON-ALGEBRAIC

B. Lindström
Dept. of Mathematics, Box 6701, S-113 85 Stockholm, Sweden

INTRODUCTION

After the definition of matroids in the 30's a lot of work
has been done on vector representations of matroids. Much less work has
been devoted to algebraic representations of matroids in the proper
sense: after the pioneer work of S. MacLane fifty years ago comes the
work of A.W. Ingleton et al. fifteen years ago. Not more than five years
ago I decided to try to solve some of the open problems in this neglected
field. Recently more people have become interested in it. I intend to
give a survey of contributions I know about.

First I recall some definitions. For a better introduction
I recommend Chapter 11 of the book of Welsh (1976).

Let F be a fixed field and K an extension of F. Elements
$e_1,\ldots,e_n$ in K are algebraically dependent over F when there is a
non-zero polynomial $p(X_1,\ldots,X_n) \in F[X]$ such that $p(e_1,\ldots,e_n) = 0$.
If E is a finite subset of K then the algebraically independent subsets of E over F give the independent sets of a matroid $M(E)$. Such
a matroid is called algebraic. The rank $r(A)$ of a subset $A \subseteq E$ in this
matroid is the transcendence degree $\mathrm{tr.d.}_F F(A)$ of the field $F(A)$
over F.

If K has finite transcendence degree over F then the
relatively algebraically closed subfields of K, containing F, form a
geometric lattice by Proposition 3.3 of Crapo & Rota (1970). The associated
combinatorial geometry is not a matroid – our matroids are finite!

ON THE ALGEBRAIC CHARACTERISTIC SET OF MATROIDS

An algebraic representation of a matroid $M(E)$ over a field
F is obtained by mapping E into an extension field K of F, $f : E \to K$,
such that a subset $A \subseteq E$ is independent in $M(E)$ if and only if
$|f(A)| = |A|$ and $f(A)$ is algebraically independent over F.

The set of characteristics of fields over which M has an algebraic representation is denoted by $\chi_A(M)$. Let $\chi_L(M)$ be the set of characteristics of fields over which M has a vector representation. Then we have $\chi_L(M) \subseteq \chi_A(M)$ by Theorem 11.2.1 of Welsh (1976). Equality does not hold e.g. when M is the non-Fano matroid of fig. 1. This matroid is algebraic over any field F (let x,y,z be algebraically independent transcendentals over F), but it is not linear over any field of characteristic 2. A matroid is called *algebraic* if $\chi_A(M)$ is non-void.

Using derivations of fields Ingleton (1971) proved

Theorem 1. If M is algebraic over a field F of characteristic 0 then M is linear over an extension field K of F.

Corollary 1.1. If $0 \in \chi_A(M)$ then $0 \in \chi_L(M)$.

Corollary 1.2. If $0 \in \chi_A(M)$ then $\chi_A(M)$ is cofinite.

The last corollary depends on a corresponding result for $\chi_L(M)$ by Rado, (4) on p. 154 in Welsh (1976).

Vámos has proved that $0 \notin \chi_L(M)$ implies that $\chi_L(M)$ is finite, (5) on p. 154 in Welsh (1976). One is tempted to guess that this is true also for the algebraic characteristic set, but this is wrong.

Consider the non-Pappus matroid depicted in fig. 2. It is non-linear, hence $0 \notin \chi_A(\text{non-Pappus})$ by Corollary 1.1. There is an algebraic representation over $GF(p^2)$ for any prime p by Lindström (1986b):

$f(1) = x^p + y$, $f(2) = x$, $f(3) = x + y$, $f(4) = y + z$, $f(5) = y + \lambda z$, $f(6) = z$, $f(7) = (\lambda - 1)x^p + \lambda y + \lambda z$, $f(8) = x^p + y + z - z^p$, $f(9) = \lambda z - x$, $\lambda^p \neq \lambda$, with x,y,z algebraically independent transcendentals over $GF(p^2)$ and $\lambda \in GF(p^2)$.

Figure 1.   Figure 2.

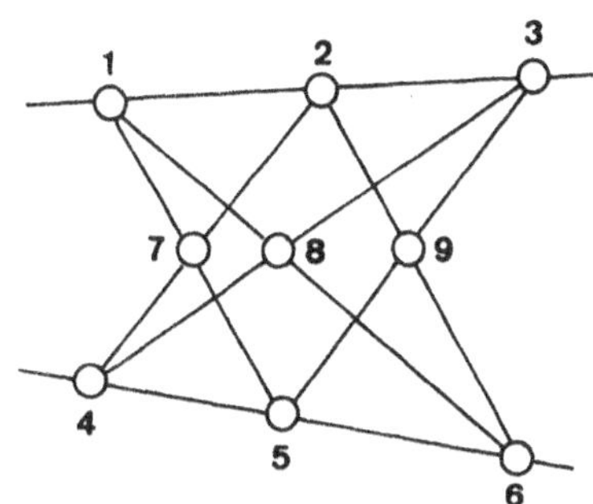

The Fano matroid is an instance (L_2) of a class of Lazarson matroids L_p for which I proved (1985a) using derivations of fields

Theorem 2. $\chi_A(L_p) = \{p\}$, where p is a prime.

G. Gordon (1985) has studied some 'sequentially unique' matroids M with $\chi_A(M)$ finite. Using derivations of fields he was able to determine $\chi_A(M_p)$ for matroids M_p of rank three with the (column) vector representation over GF(p) given by the matrix

$$\begin{pmatrix} 1 & 1 & 0 & 1 & 0 & 1 & 0 & 1 & 0 & \ldots & 1 & 0 \\ 0 & 0 & 0 & 1 & 1 & 1 & 1 & 1 & 1 & \ldots & 1 & 1 \\ 0 & 1 & 1 & 0 & 0 & 1 & 1 & 2 & 2 & \ldots & p-1 & p-1 \end{pmatrix} .$$

The matroid M_3 is depicted in fig. 3. Gordon proved the interesting

Theorem 3. $\chi_A(M_p) = \{p\}$, when p is a prime.

M. Lemos (1986) has also determined $\chi_A(M)$ for some matroids using derivations of fields. Gordon and Lemos found the first examples of matroids with $\chi_A(M)$ finite of size larger than one, Brylawski-matroids.

Lemos was mainly interested in matroids $M_p(A)$ with column vector representation (I,A) over GF(p) , where I is a unit matrix and A is a matrix of 0's and 1's. He considered special matrices A called 'algebraically admissible', e.g. matrices with a row or column of 1's and circulant matrices A_n^s of order n with s consecutive 1's in one row.

Let $_*A_n^s$ be the matrix obtained by adjoining a column of 1's to the matrix A_n^s . Lemos proved among other interesting things the following results (p is a prime)

Theorem 4. $\chi_A(M_p(A_n^p)) = \{p\}$, $2 < p < n \le 2p - 2$,
$\chi_A(M_p(_*A_n^p)) = \{p\}$, $p \nmid n$.

Figure 3.

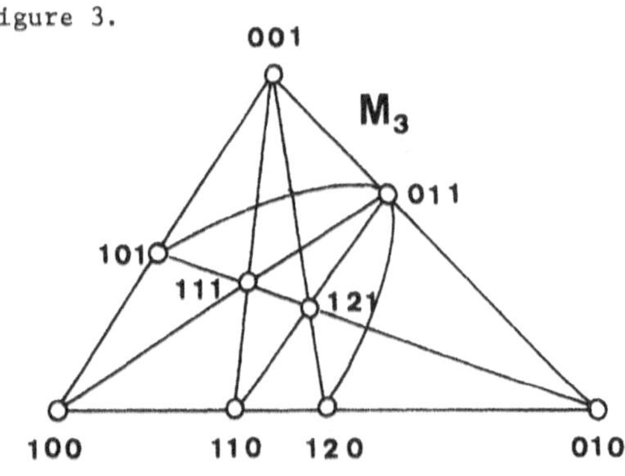

GEOMETRIC RESULTS

When Ingleton & Main (1975) proved that Vámos' cube is a non-algebraic matroid the proof depended on an algebraic lemma, which I call the Ingleton-Main lemma. In a slightly different form it is

Ingleton-Main lemma 1. Let $E = \{a_1,b_1,a_2,b_2,a_3,b_3\}$ be six transcendentals over F such that $r(E) = 4$, $r(a_i,b_i,a_j,b_j) = 3$ when $1 \leq i < j \leq 3$, and such that any three elements of E are algebraically independent over F. Then there is a transcendent c over F with $c \in \overline{F(a_i,b_i)}$ for $i = 1,2,3$ (the bar denotes algebraic closure).

It is convenient to introduce a geometric language here. If x,y,z are algebraically independent transcendentals over F, we will call the field $\overline{F(x)}$ a point, $\overline{F(x,y)}$ a line and $\overline{F(x,y,z)}$ a plane. Flats of higher rank can be defined similarly. The flats of $\overline{F(x_1,\ldots,x_n)}$ give a geometric lattice called a *full algebraic geometry* or FAG for brevity. The Ingleton-Main lemma can now be formulated as follows

Ingleton-Main lemma 2. Suppose we have 3 lines of a FAG of rank at least 4, such that any two of them are coplanar, but no all three coplanar. Then the lines will meet in a point.

With the aid of this lemma I proved (1985b) a kind of Desargues' theorem for FAG's of rank three

Theorem 5. Let (a_1,a_2,a_3) and (b_1,b_2,b_3) be two point-triples (triangles) in a plane FAG, which are in perspective from a line (i.e. corresponding sides of the triangles meet in three points on a line). Then the triangles are in perspective from a point (i.e. the lines $\overline{a_i,b_i}$ meet in a point).

Corollary 5.1. The non-Desargues matroid is non-algebraic.

Figure 4.

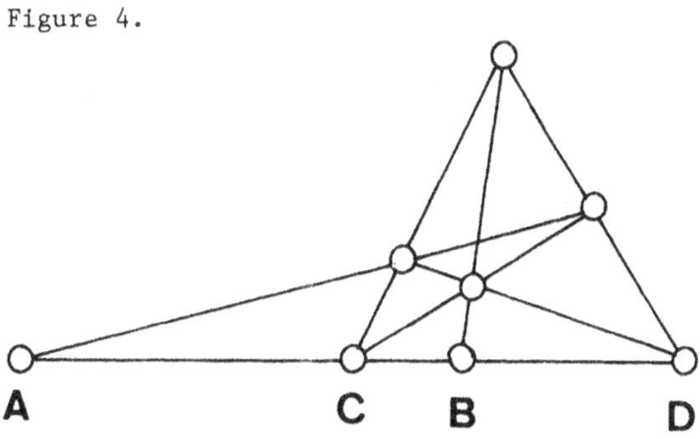

Strictly speaking, I proved the converse of Desargues' theorem for FAG's. Note that two lines in a plane FAG do not always meet. Consider e.g. the lines $\overline{x,y}$ and $\overline{z,xz+y}$ in the $\overline{x,y.z}$-plane. Therefore FAG's are not projective geometries.

In Lindström (1986a) I defined harmonic pairs of points on a line in a plane FAG. The points A and B (not necessarily distinct) are called *harmonic conjugates* with respect to two points C,D when points exist in the plane which satisfy the picture fig. 4.

Example. It is easy to see that $\overline{x+y}$ and $\overline{x-y}$ are harmonic conjugates with respect to $\overline{x}$ and $\overline{y}$.

A theorem on the fourth harmonic point was proved

Theorem 6. Assume that all points of fig. 4 exist in a plane FAG, except possibly A. Then A exists and it is unique when B,C,D are given.

D.M. Evans (1987) proves the existence of points without a harmonic conjugate when the characteristic is 0 .

Let M_n be a matroid of rank 3 with the elements A_0, B_0, A_1, B_1,...,A_{n-1}, B_{n-1}, C_0, C_1, D containing the following hyperplanes (lines) of size at least three: $\{A_0, A_1,...,A_{n-1}, D\}$, $\{B_0, B_1,...,B_{n-1}, D\}$, $\{C_0, C_1, D\}$, $\{A_i, B_i, C_0\}$, $\{A_i, B_{i+1}, C_1\}$, $0 \le i \le n-1$, $B_n = B_0$. The matroid M_3 was depicted in fig. 3. The matroid M_4 is depicted in fig. 5. With the aid of Theorem 6 I could prove (1988a) the theorem

Theorem 7. The matroids M_{2k} $(k \ge 2)$ are non-algebraic.

I guess that M_n has an algebraic representation only when n is a prime in which case we have $\chi_A(M_p) = \{p\}$ by Theorem 3.

Figure 5.

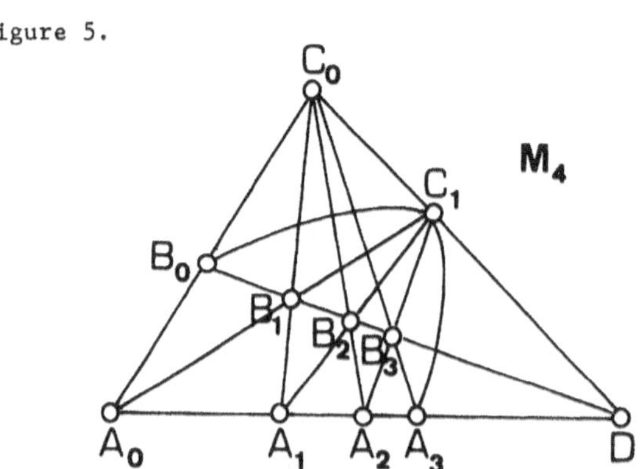

The following conjecture, if true, can be   useful.

Conjecture. Let A,B,C,D be four points of a plane FAG such that no three of them are collinear. Assume that the lines AC and BD meet and also that the lines AD and BC meet. Then the lines AB and CD will meet.

I can prove this conjecture for fields of characteristic 0 using a theorem of Ash & Rosenthal (1986). I am indebted to D.M. Evans for this reference. Evans (1987) applies this powerful tool to prove that two plane FAG's over algebraically closed fields F and F' of characteristic 0 are isomorphic as geometries only if $F \simeq F'$.

In Lindström (1988b) I studied a class of projective geometries embedded in FAG's of prime characteristic p . In this paper I call a polynomial $P(X) \in F[X]$ a p-*polynomial* if it is a sum of terms aX^{p^m} with $a \in F$ and $m \geq 0$. If $P_1,\ldots,P_n$ are p-polynomials, not all 0 , then $P_1(x_1)+\ldots+P_n(x_n)$ represents a point in the FAG of $\overline{F(x_1,\ldots,x_n)}$. The main result of Lindström (1988b) is that these points give a projective geometry. Since this geometry is Desarguesian by Theorem 5, it follows that it can be coordinatized by a skew-field. This skew-field is a ring of fractions. I will discuss it briefly.

The p-polynomials in $F[X]$ give a ring if we define the sum of two elements as the ordinary sum $P(X) + Q(X)$ and the product P o Q by substitution $P(Q(X))$. It can be verified that the ring has no zero-divisors and that it satisfies the left Ore-condition: given two p-polynomials P and Q there are p-polynomials R and S such that R o P = S o Q . It is a classic result by Ore (1931) that a ring with these properties can be embedded in a skew-field, a ring of fractions.

If $F \neq GF(p)$ then we can find $\lambda \in F$ such that $\lambda^p \neq \lambda$. In this case the ring will be non-commutative. For if we choose $P(X) = \lambda X$ and $Q(X) = X^p$, we have P o Q $\neq$ Q o P . It is a classic result in projective geometry that the coordinatizing skew-field is commutative if and only if Pappus' theorem holds. In our case it does not hold. This implies the existence of an algebraic representation of the non-Pappus matroid. Of course, we have already seen such an algebraic representation - and it was p-polynomial! The above ring-theory gives a 'natural explanation'. But this does not explain the representation in Lindström (1983)!

GENERALIZATIONS OF THE INGLETON-MAIN LEMMA

A first generalization of the Ingleton-Main lemma, the series reduction theorem, was found by Dress & Lovász (1985).

Theorem 8. Let $F \subseteq K$ be algebraically closed fields and let $S \subseteq A \subseteq K$, where A is finite and S is in series in the algebraic matroid of A over F. Then there is a $\beta \in K$ such that, for each subset $T \subseteq A - S$, the set $T \cup \{\beta\}$ is a circuit if and only if $T \cup S$ is a circuit of the algebraic matroid on A. (S is in series in A if contracting $A - S$, the set S becomes a circuit).

A stronger result appeared in a second version of their paper:

Theorem 9. Let $A(K/F)$ be a FAG over F. Let U and V be two flats of $A(K/F)$. Then there exists a flat $T = T(U,V)$ in U such that for each flat $W \subseteq U$, we have $T \subseteq W$ if and only if $r(W \cup V) - r(W) = r(U \cup V) - r(U)$.

In the time between the two versions of the Dress-Lovász paper I proved (1985c) a special case, n = 1, of the following 'conjecture':

Corollary 9.1. Let a, b, c be three flats of a FAG and $n \geq 1$ such that $r(a \vee b \vee c) = r \geq 3n$, $r(a \vee b) = r(a \vee c) = r(b \vee c) = r - n$ and $r(a) = r(b) = r(c) = r - 2n$. Then $a \wedge b = a \wedge c = b \wedge c$ and the rank of this flat is $r - 3n$.

The property of FAG's expressed by Theorem 9 was formalized by Björner & Lovász (1987). These authors define a class of semimodular lattices called pseudomodular and prove the following generalization.

Theorem 10. A semimodular lattice is pseudomodular if and only if it has the following property. If $a,b,c \in L$ have $r(a \vee c) - r(a) = r(b \vee c) - r(b) = r(a \vee b \vee c) - r(a \vee b)$, then $r((a \vee c) \wedge (b \vee c)) - r(a \wedge b) = r(a \vee c) - r(a)$.

Theorem 8 was used to show that the minimax theorem for matroid matching, proved for linear matroids by Lovász, remains valid for algebraic matroids.

I obtained (1985c) an infinite class of non-algebraic matroids generalizing the Vámos cube. Corollary 9.1 implies that their duals are non-algebraic.

Björner has constructed continuous geometries from partition lattices and observes that a similar construction is possible using FAG's thanks to the pseudomodular property.

ON A CONJECTURE BY M.J. PIFF

M.J. Piff proved the following result in his thesis (1972).

Theorem 11. If a matroid M is algebraic over a field F, then M is algebraic over a transcendent extension $P(t_1,\ldots,t_m)$ of the prime field P of F. P is Q or $GF(p)$.

The proof depends on the lemma:

Lemma. If a matroid M is algebraic over a field $F(\alpha)$ and α is algebraic over F, then M is algebraic over F.

Piff conjectured the following result, which is proved by Shameeva (1985) and Lindström (1987). Unfortunately both proofs are incomplete, but I have recently found an elementary proof.

Theorem 12. Let a matroid M be algebraic over the field $F(t)$ with t transcendent over F. Then M is algebraic over F.

Corollary 12.1. If a matroid M is algebraic over a field F, then M is algebraic over the prime field P of F.

SOME OPEN PROBLEMS

Finally I would like to mention som open problems.

- Which characteristic sets $\chi_A(M)$ are possible?
- If a matroid M is algebraic over Q, does it follow that M is algebraic over all prime fields?
- Does $\chi_A(M) = \chi_A(M^*)$ hold for every matroid M and its dual?
- Prove the conjectures in the section 'Geometric results'.
- Which are the maximal projective geometries in FAG's?

REFERENCES

Ash, C.J. & Rosenthal, J.W. (1986). Intersections of algebraically closed fields. Ann. Pure Appl. Logic, <u>30</u>, 103-19.

Björner, A. & Lovász, L. (1987). Pseudomodular lattices and continuous matroids. Dep. of Math., Univ. of Stockholm, Reports 1987-No 4.

Crapo, H. & Rota, G.C. (1970). Combinatorial Geometries. Cambridge, Mass.: M.I.T. Press.

Dress, A. & Lovász, L. (1985). On some combinatorial properties of algebraic matroids. Institut für Ökonometrie und Operations Research, Bonn, Report.

Evans, D.M. (1987). Geometries from algebraically closed fields. Letter.
Gordon, G. (1985). Algebraic characteristic sets of matroids. Preprint.
Ingleton, A.W. (1971). Representations of matroids. In Combinatorial
        Mathematics and Its Applications, ed. D.J.A. Welsh, pp. 149-69.
        London & New York: Adademic Press.
Ingleton, A.W. & Main, R.A. (1975). Non-algebraic matroids exist. Bull.
        London Math. Soc., 7, 144-6.
Lemos, M. (1986). An extension of Lindström's result about characteristic
        sets of matroids. Preprint.
Lindström, B. (1983). The non-Pappus matroid is algebraic. Ars Comb.,
        16B, 95-6.
Lindström, B. (1984). A simple non-algebraic matroid of rank three.
        Utilitas Math., 25, 95-7.
Lindström, B. (1985a). On the algebraic characteristic set for a class of
        matroids. Proc. Amer. Math. Soc., 95, 147-51.
Lindström, B. (1985b). A desarguesian theorem for algebraic combinatorial
        geometries. Combinatorica, 5, 237-9.
Lindström, B. (1985c). A generalization of the Ingleton-Main lemma and
        a class of non-algebraic matroids. Preprint.
Lindström, B. (1986a). On harmonic conjugates in algebraic combinatorial
        geometries. Europ. J. Comb., 7, 259-62.
Lindström, B. (1986b). The non-Pappus matroid is algebraic over any finite
        field. Utilitas Math., 30, 53-5.
Lindström, B. (1986c). A non-linear algebraic matroid with infinite
        characteristic set. Discrete Math., 59, 319-20.
Lindström, B. (1987). A reduction of algebraic representations of
        matroids, Proc. Amer. Math. Soc., 100, 388-9.
Lindström, B. (1988a). A class of non-algebraic matroids of rank three.
        Geometriae Dedicata.
Lindström, B. (1988b). On p-polynomial representations of projective
        geometries in algebraic combinatorial geometries.
        Math. Scand.
MacLane, S. (1938). A lattice formulation for transcendence degrees and
        p-bases. Duke Math. J., 4, 455-68.
Ore, O. (1931). Linear equations in non-commutative fields. Ann. Math. II.
        Ser., 32, 463-77.
Piff, M.J. (1972). Some problems in combinatorial theory. Thesis. Oxford.
Shameeva, O.V. (1985). Algebraic representability of matroids. Vestnik
        Moskovskogo Univ. Matem., 40, no. 4, 29-32.
Welsh, D.J.A. (1976). Matroid Theory. London: Academic Press.

Algebraic Properties of a General Convolution

I.G. Rosenberg, Mathématiques et statistique, Université de Montréal

C.P. 6128, Succ. "A" Montreeal Qué. H3C 3J7

Abstract

We study certain basic properties of algebras with convolution defined on maps from a partial groupoid into a ring-like structure. We impose fairly general conditions on a set D of such maps so that the convolution is defined on D and D is closed under it. This encompasses a fair number of structures studied in various contexts (e.g. matrices, polynomials, power series, semigroup rings and incidence algebras) of which many have combinatorial applications. We study conditions for standard properties of convolution like commutativity, associativity, idempotence, distributivity, existence of an identity and cancellation and conditions for the convolution algebra to be a semiring, dioid, ring, bisemilattice, lattice and boolean algebra.

1. Introduction

This paper develops some basic algebraic properties of convolutive algebras. The fairly general convolution is defined on a set D of maps from a partial groupoid $< B; \circ >$ into $< A; +,\cdot,0 >$ where $< A; +, 0 >$ is an abelian monoid and 0 is annihilating for the multiplication . For $f,g \in D$ the convolution $h: = f * g$ is the map from B to A defined by setting

$$h(z): = \sum_{x \circ y = z} f(x)g(y)$$

for all $z \in B$ where we sum over all $(x,y) \in B^2$ such that $x \circ y$ is defined and equal z.

We impose conditions on D so that the above sum is finite and strive to make this definition as general as possible. Such convolution encompasses a fair number of structures studied in various contexts including some familiar ones as matrices, polynomials, power series, semigroup rings and incidence algebras. Other examples are given in § 2 but certainly the list is not complete. The main part of the paper is devoted to the study of necessary and sufficient conditions for certain standard properties of convolution like commutativity, associativity, idempotence, distributivity, existence of an identity, cancellation etc. In most cases we get reasonable, although sometimes complicated, conditions. Since our definition is quite general, a certain amount of mostly routine work is involved. Somewhat surprisingly the conditions for the existence of convolutive identity (as well for cancellation) are not transparent in general and so must be postulated separately in specific cases. For this reason we have not attempted here to study inversible elements (Möbius inversion) which seem quite interesting.

Denoting by + the pointwise addition we characterize $\underline{A}$ and $\underline{B}$ such that <D ; + , * > is a semiring, ring, boolean algebra, bisemilattice, lattice etc. To keep the paper algebraic, well focused and within reasonable bounds, we have not attempted to study convergence or apply topological methods.

Special instances of our convolution play a certain role in combinatorics and related fields, e.g. in enumeration (power series and Möbius inversion), orders (Möbius inversion), paths in graphs (dioids), combinatorial number theory (Möbius inversion), combinatorial optimization (dioids of Gondran & Minoux (1979)), designs (Graver & Jurkat (1972,1973) cf also Deza & Rosenberg (1986)), automata (power series, Salomaa & Soittola (1978)) and categories, cf Leroux (1976), Content (1977), Content et al (1980), Joyal (1981) and Labelle (1983).

A further generalization of convolution is in Deza & Rosenberg (1986) where $x \circ y$ is a subset of B for all $x, y \in B$ (i.e. $\circ$ is a partial multi-,hyper- or polygroupoid) and the summation is over all $(x,y) \in B^2$ with $z \in x \circ y$. (The paper also contains indications on applications to generalized designs.) In spite of the existence of this more general concept, it seems that the present one should be explored first since it is a common umbrella for many existing structures and leads to sufficiently complex problems. The main purpose of this paper is to draw attention to this concept. The obvious next step is to look at specific results in semigroup rings, power series etc. and to try finding conditions under which these extend to generalized convolution, a long range project given the size of the literature.

2. Preliminaries

2.1   Let $\underline{A} = <A ; +, \cdot, 0>$ be a set A of cardinality > 1 with two binary operations (groupoids) $+$ and $\cdot$ and an element $0 \in A$ such that $<A ; +, 0>$ is an abelian monoid (i.e. $+$ is commutative and associative and 0 its neutral element) and $a0 = 0a = 0$ holds for all $a \in A$. Further let $\underline{B} = <B ; \circ>$ be a partial groupoid (i.e. $(x_1, x_2) \to x_1 \circ x_2$ is a map from a subset N of B^2 into B). To avoid trivial cases we assume that $|A| > 1$ and $N \neq \phi$. For $X, Y \subseteq B$ put

$$X \bullet Y := \{ x \bullet y : (x,y) \in N \cap (X \times Y) \}.$$

Finally let $\underline{C}$ be a family of subsets of B such that:

(i) $\underline{C}$ is an order ideal in $(2^B, \subseteq)$ (i.e. $X \cup Y \in \underline{C}$ whenever $X, Y \in \underline{C}$ and also $\underline{C}$ is hereditary in the sense that $Z \in \underline{C}$ whenever $Z \subseteq X \in \underline{C}$) containing all singletons from B, and

(ii)   For all $X, Y \in \underline{C}$ the set $X \circ Y$ belongs to $\underline{C}$ and $\{ (x,y) \in N \cap (X \times Y) : x \circ y = b \}$ is finite for each $b \in B$.

As usual, the set of maps $B \to A$ is denoted by A^B and for $f \in A^B$ the support supp f of f is the set $\{ b \in B : f(b) \neq 0 \}$. We are interested in the set

$$D := D_{\underline{ABC}} := \{ f \in A^B : \operatorname{supp} f \in \underline{C} \} .$$

The operation $+$ on D is the usual componentwise addition : given f, $g \in D$ the map $f + g$ is defined by setting $(f+g)(b) := f(b) + g(b)$ for all $b \in B$. The condition (i) guarantees that D is closed under $+$. The convolution is the following binary operation $*$ on D : given $f, g \in D$ put

$$(f * g)(b) := \sum_{x \circ y = b} f(x)g(y) \qquad (1)$$

for every $b \in B$ (where the summation is understood to be over all pairs $(x,y) \in N$ such that $x \circ y = b$) . In view of (ii) the sum in (1) is finite and so $(f*g)(b)$ is a well defined element of A (by the usual convention it is 0 if the summation is over the empty set). Morever, by (1) and (ii)   supp $(f*g) \subseteq \{ x \circ y : (x,y) \in N \cap (\operatorname{supp} f \times \operatorname{supp} g) \} \in \underline{C}$ and so by (i) also supp $(f*g) \in \underline{C}$ proving $f*g \in D$. Let $\underline{0}$ denote the constant map from B into A with the value 0 . Clearly $<D; +, \underline{0} >$ is an abelian monoid and $\underline{0} * f = f * \underline{0} = \underline{0}$ for all $f \in D$. We call $\underline{D} := < D; +, * , \underline{0} >$ a <u>convolutive algebra</u> . In this paper we study certain properties of convolutive algebras. Note that D is closed under the componentwise (or direct) product as well; however, this operation will not be used in this paper.

The symbol F_B denotes the family of all finite subsets of B. For $\underline{C} := F_B$ the corresponding $\underline{D}$ is denoted by

$A\,\underline{B}_{fin}$ and called a <u>restricted convolutive algebra</u>. Clearly $A\,\underline{B}_{fin}$ is a convolutive subalgebra of each convolutive algebra (on $\underline{A}$ and $\underline{B}$).

2.2   Example. Put $N := \{ (b,b) : b \in B \}$ and $b \circ b := b$ for all $b \in B$.
Then $*$ is the componentwise multiplication (i.e. $(f*g)(b) = f(b) \cdot g(b)$ for all
$b \in B$). For $C := 2^B$ the structure $\underline{D}$ is the full direct power $\underline{A}^B$. Choosing
$\underline{C} := F_B$ the structure $\underline{D}$ is the restricted or weak direct power.

2.3   Example. Let n be a positive integer, $\underline{n} := \{1,2,...,n\}$ and $B := \underline{n}^2$. Put
$N := \{ ((x,y),(y,z)) : x,y,z \in \underline{n} \}$ and let $(x,y)\circ(y,z) := (x,z)$. Put $\underline{C} := 2^B$. If
$\underline{A}$ is a ring then $\underline{D}$ is isomorphic to the standard ring of all $n \times n$ matrices over $\underline{A}$.
For a lattice $\underline{A}$ the matrix lattice $\underline{D}$ has been considered in switching theory (cf
(16)), automata and fuzzy set theory. There is also an obvious extension to infinite
matrices: replace $\underline{n}$ by a set I and let $\underline{C}$ consist of $X \subsetneq I^2$ such that for all $i \in I$
both sets $\{ x \in I : (x,i) \in X \}$ and $\{ y \in I : (i,y) \in X \}$ are finite.

2.4   Example. Let $B = \mathbb{N} := \{ 0,1,... \}$ (the set of non-negative integers),
Put $N := \mathbb{N}^2$, let $x \circ y$ be the sum of x and y and let $\underline{C} := F_{\mathbb{N}}$. If $\underline{A}$ is a ring
then $\underline{D}$ is isomorphic to the ring $\underline{A}[x]$ of all polynomials in one indeterminate over
$\underline{A}$ (the isomorphism sends $f \in D$ into $\sum_{i \in \mathbb{N}} f(i)\, x^i)$. To obtain the ring
$\underline{A}[x_1,...,x_n]$ in several commuting indeterminates it suffices to choose $B := \mathbb{N}^n$, N
$:= B^2$ and $(b_1',...,b_n') \circ (b_1'',...,b_n'') := (b_1' + b_1'',...,b_n' + b_n'')$.

     For the next example we need the foolowing:

2.5   Let $\leq$ be a linear order (i.e. total order or chain) on B such that (x,y) ,
$(x',y') \in N$, $x \leq x'$, $y \leq y'$ implies $x \circ y \leq x' \circ y'$ with equality exactly if $x=x'$
and $y=y'$. Let ξ be an infinite cardinal and let W_ξ denote the family of all
well-orderd subsets of B of cardinality at most ξ (a set $X \subseteq B$ is well-ordered if

each nonempty subset Y of X has a least element in $\leq$). We need :

2.6   Lemma $\underline{C} := W_\xi$ *satisfies* (i) *and* (ii) *in* 2.1

Proof : Clearly W_ξ is closed under union, is hereditary and contains all singletons, hence $\underline{C}$ satisfies (i). To prove (ii) first note that if $x \circ y = x' \circ y'$ (this notation tacitly assumes both $(x,y) \in N$ and $(x',y') \in N$) then

$$x < x' \Rightarrow y > y' , \quad y < y' \Rightarrow x > x' . \tag{2}$$

Indeed, if, say, $x < x'$ and $y \leq y'$ then $x \circ y \leq x' \circ y'$ but here by assumption equality cannot hold since $x \neq x'$. Suppose there are $X, Y \in W_\xi$ and $b \in B$ such that there is a countably infinite set $\{(x_i, y_i) \in N : i \in \mathbb{N}\}$ satisfying $x_i \in X$, $y_i \in Y$ an $x_i \circ y_i = b$ for all $i \in \mathbb{N}$. Since X is well-ordered, we can arrange the notation so that $x_0 \leq x_1 \leq \dots$. Moreover, all the inequalities are strict. Indeed, if $x_i = x_{i+1}$ then $y_i \neq y_{i+1}$. However, by (2) (since $x_i \circ y_i = x_{i+1} \circ y_{i+1} = b$) we get either $x_i < x_{i+1}$ or $x_i > x_{i+1}$ contrary to $x_i = x_{i+1}$. By the same token we get $y_0 > y_1 >$ … in contradiction to the fact that all y_i belong to the well-ordered seet Y.

To finish the proof set $Z := \{x \circ y : (x,y) \in N \cap (X \times Y)\}$. Clearly $|Z| \leq |X| \cdot |Y| \leq \xi^2 = \xi$. We must prove that Z is well-ordered. Suppose there are $(x_i, y_i) \in N \cap (X \times Y)$ such tat $x_0 \circ y_0 > x_1 \circ y_1 > \dots$. Let x_{p_1} denote the least element of the sequence $< x_0, x_1, \dots, >$. For each $j > p_1$ we have $y_{p_1} > y_j$ (because otherwise $x_{p_1} \circ y_{p_1} \leq x_j \circ y_j$). Let y_{p_2} be the least element of $< y_{p_1+1}$, $y_{p_1+2}, \dots >$. By the same token we have $x_{p_2} > x_j$ for each $j > p_2$. Let x_{p_3} be the least element of $< x_{p_2+1}, x_{p_2+2}, \dots >$. Then $y_{p_3} > y_j$ for each $j > p_3$. Continuing in this fashion we again construct an infinite descending chain $y_{p_1} > y_{p_2} > \dots$ in the well-ordered set Y. $\square$

Returning back to our examples we have:

2.7 Example. Let $B := <\mathbb{N}, +>$ be the additive semigroup of non-negative integers with the natural order and let $\underline{C} = W_{\aleph_0}$. Clearly $\underline{C}$ is the set of all subsets of $\mathbb{N}$.

Moreover, the addition in $\mathbb{N}$ is order preserving and $x \leq x'$, $y \leq y'$ and $x + y = x' + y'$ implies $x = x'$ and $y = y'$. Thus by 2.6 the algebra $\underline{D}$ is well defined. For $\underline{A}$ a ring it may be easily seen that $\underline{D}$ is isomorphic to the ring $\underline{A}[[x]]$ of formal power series over $\underline{A}$ (the isomorphism sends $f \in A^{\mathbb{N}}$ into $\Sigma_{n \in \mathbb{N}} f(n) x^n$). If we take $B = <\mathbb{N}^n, +>$ (with the vector, i.e. componentwise, addition) we obtain the ring $\underline{A}[[x_1,...,x_n]]$ of formal power series in n commuting indeterminates over $\underline{A}$.

2.8 Example. Let $\underline{B} := <\mathbb{Z}, +>$ (the additive group of integers) with the natural order and let $\underline{C} := W_{\aleph_0}$. By 2.6 we can form $\underline{D}$ which is isomorphic to set

$$\{ \sum_{i \in Y} a_i x^i : Y \subseteq \mathbb{Z} \text{ has a least element} \}$$

with the usual multiplication.

2.9 Example. Let $\leq$ be a linear order on B satisfying the condition of 2.5, let ξ be an infinite cardinal and let $\underline{R_\xi} := \{ X \in W_\xi : |X| \geq \aleph_0 \Rightarrow X \text{ is unbounded}\}$ (i.e. has no upper bound). Clearly $\underline{R_\xi}$ satisfies (i) and, as a subset of W_ξ, the condition (ii) as well. Hence we can form $D_{\underline{ABR_\xi}}$ which for $\underline{B} := <\mathbb{R}; +>$ (the reals with the natural sum and order) and $\xi := \aleph_0$ was introduced by Laugwitz (1968) as a model of nonstandard arithmetic.

2.10 Example. Let S be a set, $B := S^2$, $N := \{ ((p,q), (q,r)) : p,q,r \in S \}$, and $(p,q) \circ (q,r) := (p,r)$ for all $p,q,r \in S$ (cf Example 2.3). Let $\underline{C}$ be a hereditary system of binary relations on S, containing all singletons $\{(x,y)\}$ (for $x,y \in S$), closed under union and relational product (where for $X, Y \subseteq S^2$ the product XY is

$\{(x,y) : (x,u) \in X, (u,y) \in Y$ for some $u \in S\}$) and such that the set $\{ u \in S : (p,u)$

$\in X, (u,r) \in Y\}$ is finite whenever $X,Y \in \underline{C}$ and $p,r \in S$ (in particular, for every

$X \in \underline{C}$ setting $X=Y$ we obtain that X *is locally finite*, i.e. for all $p,q \in S$ there are

only finitely many $u \in S$ with $(p,u),(u,q) \in X$). Setting $\underline{C}$ to be the set of all

subrelations of a transitive and locally finite binary relation Z and $\underline{A}$ to be the field

of reals, the structure $\underline{D}$ is the one introduced by Finch (1970). If, moreover, Z

is a locally finite partial order $\leq$ on B then $\underline{D}$ is the incidence algebra. Indeed, the

elements of the incidence algebra are the maps $f: B^2 \to A$ such that supp $f \subseteq Z$ (

meaning $f(x,y) = 0$ whenever $x \leq y$) and $(f * g)(x,y) := \Sigma_{x \leq u \leq y} f(x,u) \, g(u,y)$

for all $f,g \in \underline{D}$ and $x,y \in B$.

2.11 Example. Let $\underline{B} := \, <B; \circ>$ be a commutative semigroup (i.e. $N := B^2$ and

$\circ$ is associative), $\underline{C} := F_B$ an let $\underline{A}$ be a ring. Then $\underline{D}$ is the semigroup ring of $\underline{B}$

over $\underline{A}$. If, moreover, $\underline{B}$ is a group (not necessarily abelian) then $\underline{D}$ is a

group-ring. For $\underline{B}$ a monoid and $\underline{A}$ a dioid (cf § 4) the elements of D are called

formal power series in non-commuting variables (cf Salomaa & Soittola (1978) pp

11-15.

3. <u>Properties of convolution.</u>

3.1 It is natural to ask under what conditions the general convolution has certain

standard properties. We start with commutativity. For the proof we need the *peak*

functions $|a|_b \in A^B$ ($a \in A, b \in B$) defined by setting $|a|_b(b) := a$ and $|a|_b(x) := 0$

for all

$x \in B \setminus \{b\}$. Note that by (i) we have $\{b\} \in \underline{C}$, hence supp $|a|_b = \{b\} \in \underline{C}$ shows

$|a|_b \in D$. Recall that $\underline{B}$ is a *commutative* partial groupoid if $(y,x) \in N$ and $y \bullet x$

$= x \circ y$ whenever $(x,y) \in N$ (in other words, the multiplication table of $\circ$ is

symmetric).

3.2 Proposition. *The following conditions are equivalent for a convolution algebra:*

(A)   *the convolution is comutative,*

(B)   *the convolution of the restricted algebra is commutative,*

(C)   *the peak functions pairwise commute,*

(D)   *if $A^2 \neq \{0\}$, then both $<A;.>$ and $\underline{B}$ are commutative.*

Proof: (A) $\Rightarrow$ (B) $\Rightarrow$ (C) is evident. (C) $\Rightarrow$ (D) : Suppose $A^2 \neq \{0\}$. Let a,b $\in A$ satisfy a b $\neq 0$, let $(p,q) \in N$ and let $r := p \circ q$. By (C) we have $(\lceil a \rceil_p *$ $\lceil b \rceil_q (r) = (\lceil b \rceil_q * \lceil a \rceil_p) (r)$. Here the left side equals ab $\neq 0$, whence the right side does not vanish and so $(q,p) \in N$ and $q \circ p = r$ proving that $\underline{B}$ is commutative. Fix $(p,q) \in N$ (by assumption $N \neq \phi$). Now for arbitary a,b$\in A$ the above equation reduces to ab = ba proving (D). (D) $\Rightarrow$ (A) Direct check. $\square$

3.3 The next property studied is associativity. As usual, $\underline{B}$ is *associative* (or a *partial semigroup*) if $x \circ (y \circ z) = (x \circ y) \circ z$ holds whenever at least one side is definied (i.e. more explicitly, if $(y,z) \in N$, $(x,y \circ z) \in N$, then (i) $(x,y) \in N$, (x $\circ$ y,z) $\in N$ and (ii) $(x \circ y) \circ z = x \circ (y \circ z)$ and similarly in the other direction). We say that $\underline{B}$ is *left (right) trivially associative* if the left (right) side of the associative equality is never defined, i.e. if $(x \circ y,z) \notin N$ whenever $(x,y) \in N$ and similarly $(x,y \circ z) \notin N$ whenever $(y,z) \in N$. We say that $\underline{B}$ is *trivially associative* if $\underline{B}$ is both left and right trivially associative.

The necessary and sufficient conditions for the associativity of the convolution are formulated in terms of the following laws. Let $\underline{B}$ be a partial semigroup. For a finite subset Z of B and z$\in B$ the identity

$$\sum_{P \in Z} a_p \left(\sum_{q,r \in Z,\ p \circ q \circ r = z} b_q c_r \right) = \sum_{r \in Z} \left(\sum_{p,q \in Z, p \circ q \circ r = z} a_p b_q \right) c_r \qquad (3)$$

(where a_i, b_i, c_i are arbitrary elements of A for all $i \in Z$) is the *pseudo-distributive law implied by Z and z*. Finally put $A^2 \cdot A := \{ (ab)c : a,b,c \in A \}$ and define

$A \cdot A^2$ in a similar fashion. We have:

3.4 Proposition. *The following are equivalent:*

(A)   *the convolution is associative,*

(B)   *the convolution of the restricted convolutive algebra is associative,*

(C)   *one of the following contions holds:*

(i) $A^2 \cdot A = A \cdot A^2 = \{0\}$,

(ii) $A^2 \cdot A = \{0\}$ and $\underline{B}$ *is right trivially associative,*

(iii) $A \cdot A^2 = \{0\}$ and $\underline{B}$ *is left trivially associative,*

(iv) $\underline{B}$ *is trivially associative,*

(v) $\underline{B}$ is a partial semigroup, $<A;\cdot>$ *is a semigroup* and $\underline{A}$ *satisfies all the*

pseudo-distributive laws implied by the finite subsets and elements of $\underline{B}$.

Proof: $(A) \Rightarrow (B)$ is trivial. $(B) \Rightarrow (C)$. For $a,b,c \in A$ and $p,q,r \in B$ the

associative law holds for the peak functions $\ulcorner a \urcorner_p, \ulcorner b \urcorner_q, \ulcorner c \urcorner_r$ iff   for each $z \in B$

$$\sum_{xoy=z} \left(\sum_{uov=x} \ulcorner a \urcorner_p(u)\, \ulcorner b \urcorner_q (v) \right) \ulcorner c \urcorner_r (y) = \sum_{x'oy'=z} \ulcorner a \urcorner_p(x') \left(\sum_{u'ov'=y'} \ulcorner b \urcorner_q (u')\ulcorner c \urcorner_r (v') \right). \quad (4)$$

Suppose that neither of (i) - (iv) holds. Since (iv) does not hold, we may assume

by symmetry that $\underline{B}$ is not left trivially associative. We prove $A^2 \cdot A \neq \{0\}$.

Assume the contrary. Since (i) does not hold, there are $a,b,c \in A$ such that

$d := a(bc) \neq 0$. Now (ii) is not valid either, hence $z = p o (qor)$ is defined for some

$p,q,r \in B$. For our a,b,c,p,q,r and z the right side of (4) is $d \neq 0$ while the left

side vanishes. This contradiction shows $A^2 \cdot A \neq \{0\}$. Fix $a,b,c \in A$ so that

$d := (ab) c \neq 0$ and suppose $z := (p o q) o r$ is defined in $\underline{B}$. For these elements

the left side of (4) is $d \neq 0$, hence in order to have the right side nonvanishing,

$q \bullet r$ is defined, $p \circ (q \circ r)$ is defined and $p \circ (q \circ r) = z$ proving thus one half of the partial associative law. We assume that $\underline{B}$ is not left trivially associative and therefore there are $p,q,r \in B$ such that $(p \circ q) \circ r$ is defined. By the first half of the partial associative law $(p \circ q) \circ r = p \circ (q \circ r)$. From (4) it follows easily that $<A,\cdot>$ is associative. In particular $A \cdot A^2 = A^2 \cdot A \neq \{0\}$. Moreover $\underline{B}$ is not right trivially associative, hence by symmetry we obtain the other half of the partial associative law as well. The pseudo-distributive law (3) is nothing else than $f *$ $(g * h)(z) = (f * g) * h(z)$ for:

$$f := \sum_{p \in Z} \lceil a_p \rceil_p, \; g := \sum_{p \in Z} \lceil b_p \rceil_p, \; h := \sum_{p \in Z} \lceil c_p \rceil_p \;.$$

Thus (v) holds.

(C) $\Rightarrow$ (A). In the cases (i) - (iv) we have $f * (g*h) = \underline{0} = (f*g) * h$ for all $f,g,h \in D$ and so the convolution is associative. Suppose (v) holds. Let $f,g,h \in D$ and $z \in B$. The supports of f,g,h, $f*g$ and $g*h$ belong to $\underline{C}$ and so by 2.1 (ii) the set

$$S := \{ (p,q,r) \in B^3 : p \circ q \circ r = z, \; p \in \text{supp } f, \; q \in \text{supp } g, \; r \in \text{supp } r \}$$

is finite. Put $P := pr_1 S$, $Q := pr_2 S$ and $R := pr_3 S$ (where $pr_1 S := \{ p : (p,q,r) \in S \}$ and $pr_2 S$ and $pr_3 S$ are defined in a similar way). Put $Z := P \cup Q \cup R$ and

$$a_p := f(p), \; b_q := g(q), \; c_r := h(r)$$

for all $p \in P, \; q \in Q$ and $r \in R$ and

$$a_i = b_j = c_k = 0$$

for all $i \in Z \backslash P$, $j \in Z \backslash Q$ and $k \in Z \backslash R$. Now the equality $((f*g)*h)(z) = (f*(g*h))(z)$ follows from the pseudo-distributive law implied by Z and z proving that $<D; *>$ is associative. $\square$

3.5 Remark. If $\underline{A}$ is distributive (i.e. both distributive laws

$$a(b+c) = ab + ac \ , \ (a+b)c = ab + ac$$

hold for all $a,b,c \in A$) and $\underline{B}$ is a partial semigroup, then all the pseudo-distributive

laws hold and so the condition (v) in 3.4 is satisfied.

We may ask when the convolution is idempotent. We say that $\underline{B}$ is

idempotent if $(p,p) \in N$ and $p \circ p = p$ for all $p \in B$.

3.6 Proposition. *The following are equivalent :*

(A)   *the convolution is idempotent,*

(B)   *the convolution of the restricted convolutive algebra is idempotent, and*

(C)   *both* $<A; \cdot>$ *and* $\underline{B}$ *are idempotent* ,

$$(p,q) \in N \ , \ p \circ q \notin \{p,q\} \Rightarrow (p,q) \in N , \ \ p \circ q = q \circ p \qquad (*)$$

and for arbitrary $a,b \in A$

    (i) $b = b + ab$ *whenever for some* $(p,q) \in N$ *we have* a) $p \circ q = q$ *and*

       · b) *either* $(q,p) \notin N$ *or* $q \circ p \neq q,$

    (ii) $a = a + ab$ *whenever for some* $(p,q) \in N$ *we have* a) $p \circ q = p$ *and*

       b) *either* $(q,p) \notin N$ *or* $q \circ p \neq p,$

    (iii) $a = a + ab + ba$ *whenever for some* $p,q \in B, p \neq q,$ *we have*

       $(p,q),(q,p) \in N$ *and* $p \circ q = q \circ p = p,$

    (iv) $ab + ba = 0$ *whenever for some* $p,q \in B$ *we have* $(p,q),(q,p) \in N$

       while $p \circ q = q \circ p \notin \{p,q\}.$

Proof: (A) $\Rightarrow$ (B) obvious · (B) $\Rightarrow$ (C) . The idempotency of both $<A; \cdot >$ and

$\underline{B}$ follows from $\lceil a \rceil_p * \lceil a \rceil_p = \lceil a \rceil_p$. Let $a,b,c \in A$ be arbitrary. To prove (i)

suppose that for some $(p,q) \in N$ we have $p \circ q = q$ and either $(q,p) \notin N$ or $q \circ p \neq$

q . Note that $p \neq q$. Put $f := \lceil a \rceil_p + \lceil b \rceil_q$. In $f(q) = (f*f)(q)$ the left side is b

while the right side is $b + ab$ (corresponding to $q \circ q = q$ and $p \circ q = q$) and so (i)

holds. The conditions (ii) and (iii) are derived in a similar way. Suppose now that there is $(p,q) \in N$ such that $p \circ q = r \notin \{p,q\}$. Consider $f := [a]_p + [b]_q$. From

$(f*f)(r) = f(r)$ we get $S := \sum_{x \circ y = r} f(x) \, f(y) = 0$. For our f it suffices to consider

only the pairs $(x,y) \in \{ (p,q), (q,p) \} \cap N$ because $p \circ p = p \neq r \neq q = q \circ q$. If

$(q,p) \notin N$ or $q \circ p \neq r$ we have $S = f(p) \, f(q) = ab = 0$ By idempotency

$a = a^2 = 0$ for all $a \in A$, i.e. $A = \{0\}$ proving (*). if $(q,p) \in N$

and $q \circ p = r$ we get $S = ab + ba = 0$ proving (iv) and thus (C). (C) $\Rightarrow$

(A). Let $f \in D$ and $p \in B$. As $R := \text{supp } f \in \underline{C}$ the set $F := \{ (x,y) \in N \cap R^2 :$

$x \circ y = p \}$ is finite. Let $P := pr_1 F \cup pr_2 F$ denote the set of all $b \in B$ appearing in

F. Write $P := \{q_1,...,q_k\}$ where $q_1 = p$ if $f(p) \neq 0$ and for $m = 1,...,k$ put F_m

$:= F \cap \{q_1,...,q_m\}^2$. By induction on $m = 1,.., k$ we prove

$$\sum_{(x,y) \in F_m} f(x) \, f(y) \; = \; f(p). \qquad\qquad (5_m)$$

Indeed, if $F_1 = \{ (p,p) \}$ then (5_1) reduces to $f(p) \, f(p) = f(p)$ which holds since

$<A ; \cdot >$ is idempotent. If $F_1 = \phi$, then $f(p) = 0$ and both sides of (5_1) vanish.

Suppose that $1 \leq m < k$ and (5_m) holds. Put $a := f(p)$. Let $(x,y) \in F_{m+1} \setminus F_m$.

Now (x,y) satiffies one of the assumptions of (i) - (v) (for example, if $x = p$ and

$(y,x) \in F_{m+1}$, then we are in the case (iii)). Note that in the cases (i) and (ii) we

have $a + f(x) \, f(y) = a$, in the case (iii) we have $a + f(x) \, f(y) + f(y) \, f(x) = a$, in the

case (iv) we have $f(x) \, f(y) = 0$ and in the case (v) $f(x) \, f(y) + f(y) \, f(x) = 0$.

Continuing in this fashion we get (5_{m+1}). This concludes the induction; in

particular, (5_k) proves $f*f(p) = f(p)$. $\square$

3.7 Corollary. *Let* $\underline{A}$ *satisfy* $a = a+x \Leftrightarrow x = 0$ *for all* $a,x \in A$ *and let* $N \neq$

$\{<x,x> : x \in B \}$. *Then the convolution is idempotent if and only if both*

$<A ; \cdot >$ *and* $\underline{B}$ *are idempotent,* $\underline{B}$ *is commutative and* $a + a = ab + ba = 0$ *hold*

for all $a,b \in A$.

Proof: Each of the assumptions in (i) , (ii) and (iv) in 3.6 yields $ab = 0$. In

particular, $a^2 = 0$ leading (by idempotency) to the contradiction $a = a^2 = 0$ for all

$a \in A$. Thus $\underline{B}$ is commutative and from (iii)   it follows that $ab + ba = 0$ holds for all $a,b \in A$; in particular, $a + a = a^2 + a^2 = 0$. $\square$

We may ask when the convolution distributes over the addition.

3.8 Proposition. *The following are equivalent :*

(A)   *the left (right) distributive law holds in* $\underline{D}$,

(B)   *the left (right) distributive law holds in the restricted convolutive algebra,*

(C)   *the left (right) distributive law holds in* $\underline{A}$.

Proof: (A) $\Rightarrow$ (B) is evident. (B) $\Rightarrow$ (C). Let $(p,q) \in N$, $p \circ q = r$ and $a,b,c \in A$. From $\lceil a \rceil_p * (\lceil b \rceil_q + \lceil c \rceil_q) = (\lceil a \rceil_p * \lceil b \rceil_q) + (\lceil a \rceil_p * \lceil c \rceil_q)$ at r we get $a(b+c) = ab+ac)$. The right distributive law is obtained in a similar fashion. (C) $\Rightarrow$ (A). Direct check. $\square$

Now we can answer easily the question: under what conditions is $\underline{D}$ a ring? We start with a related structure. Call $<S ; + , \cdot >$ *distributive* if both distributive laws $a(b+c) = ab + ac$ and $(a + b)c = ac + bc$ hold in S. Morever, $<S ; + , \cdot >$ is a *semiring* if it is distributive and both $<S ; + >$ and $<S ; \cdot >$ are semigroups (cf. Weinert (1984)).

3.9 Corollary. *The following conditions are equivalent for a convolutive algebra:*

(A)   *the algebra is a semiring,*

(B)   *the restricted algebra is a semiring*

(C)   *either* a) $\underline{A}$ *is distributive and at least one of the conditions* (i) - (iv) *of* Prop. 3.4 *holds or* b) $\underline{B}$ *is a partial semigroup and* $\underline{A}$ *a semiring.*

Proof: Apply Propositions 3.4 and 3.8 and note that for $\underline{A}$ semiring by Remark 3.5 all the pseudo-distributive laws hold. $\square$

If $\underline{D}$ is semiring, then $\underline{0}$ is an absorbing or annihilating zero of $\underline{D}$ (which need not exist in a general semiring). Corollary   3.9 may be extended to semi-nearring (only the right distributivity is assumed) but the condition (C) is more complicated. As usual, $\underline{R}$; = $< R; +, \cdot, 0 >$ is a *ring*   if $\underline{R}$ is distributive and $<R ; +, 0 >$ an abelian group.

3.10 Corollary. *The foolowing conditons are equivalent for a convolutive algebra :*

(A)   *the algebra is a ring,*

(B)   *the restricted algebra is a ring,*

(C) $\underline{A}$ *is a ring.*

Proof: Note that $< A ; +, 0 >$ is an abelian group if $\underline{D}$ is a ring. The assertion is a direct consequence of 3.8. $\square$

The conditions for $\underline{D}$ to be an associative or commutative ring can easily be obtained by combining 3.4 or 3.2 with 3.10. Clearly $\underline{D}$ is the zero ring if and only if $A^2 = \{0\}$. The existence of identity is discussed in § 4.

A *semilattice* is a commutative idempotent semigroup. We say that $<S ; + \cdot >$ is a *bisemillattice* if both $<S ; + >$ and $<S ; \cdot >$ are semilattices (cf Romanowska (1982, 1983). We start with the following lemma. Denote $a + a$ by 2a.

3.11 Lemma. $<D ; * >$ *is both commutative and idempotent if and only if* (i)   *both* $<A ; \cdot >$ *and* $\underline{B}$ *are commutative and idempotent and* (ii) *for all* $a,b \in A$ *we have* 1) $a = a + 2ab$ *whenever* $p \circ q \in \{ p,q \}$ *for some* $(p,q) \in N, p \neq q$ *and* 2) $2 ab = 0$ *whenever* $p \circ q \notin \{ p,q \}$ *for some* $(p,q) \in N$.

Proff: Necessity. By Proposition 3.6 both $<A ; \cdot >$ and $\underline{B}$ are idempotent. It follows that $A^2 \neq \{0\}$ and by Propositon 3.2 both $<A ; \cdot >$ and $\underline{B}$ are commutative. Note that neither of the conditions (i) and (ii) from 3.6 applies. Suppose there is $(p,q) \in N$ such that $p \circ q \notin \{p,q\}$. From (iv) and the commutativity we obtain 2). Similarly, if there is $(p,q) \in N$ with $p \neq q$ and $p \circ q \in \{p,q\}$ from (iii) we get 1).

Sufficiency. Propositions 3.2 an 3.6. $\square$

3.12 Corollary. *Both $<D; + >$ and $<D ; *>$ are commutative and idempotent if and only if* (i) *$<A; + >, <A ; \cdot >$ and $\underline{B}$ are all three commutative and idempotent and* (ii) *$p \circ q \in \{p,q\}$ for all $(p,q) \in N$ and $a = a + ab$ for all $a,b \in A$ whenever there is $(p,q) \in N$ with $p \neq q$.*

Proof: Necessity. The idempotency of $<D; + >$ implies the idempotency of $<A; +>$. Suppose there is $(p,q) \in N$ with $p \circ q \notin \{p,q\}$. By 2) in Lemma 3.11 we have $a = a^2 = 2a^2 = 0$ for all $a \in A$ in contradiction to $|A| > 1$. In 1) we have $a = a + 2ab = a + ab$.

Sufficiency. Lemma 3.11. $\square$

We turn to bisemilattices. As usual, put $p \leq q$ if $(p,q) \in N$ and $p \circ q = q$. We say that $\leq$ an *oriented forest* if it is an order in which every subset bounded from above is a chain (a linearly or totally ordered subset). Let $m_{\leq}$ denote the least cardinal exceeding the length of every finite chain in $\leq$. We have:

3.13 Proposition. *The folllowing conditions are equivalent for a convolutive algebra:*

(A)   *the convolutive algebra is a bisemilattice,*

(B)   *the restricted convolutive algebra is a bisemilattice,*

(C)   (i)   *$\underline{B}$ is a partial semilattice, $p \circ q \in \{p,q\}$ for all $(p,q) \in N$ and $\leq$ is* *an oriented forest, and*

(ii) $\underline{A}$ *is a bisemilattice satisfying* a) $a = a + ab$ *for all* $a,b \in A$

whenever there is $(p,q) \in N$ *with* $p \neq q$ *and* b)

$$a \left(\sum_{i=1}^{k} \sum_{j=1}^{k} b_i c_j \right) = \sum_{j=1}^{k} \left(\sum_{i=1}^{k} ab_i \right) c_j \qquad (6)$$

holds for all $k < m_{\leq}$ *and* $a, b_1, ..., b_k, c_1, ..., c_k \in A.$

Proof: $(A) \Rightarrow (B)$ is evident. $(B) \Rightarrow (C)$. By Proposition 3.4 and Corollary 3.12 we know that $\underline{B}$ is a partial semilattice and that $p \circ q \in \{p,q\}$ for all $(p,q) \in N$. The relation $\leq$ is clearly reflexive (as $(p,p) \in N$ and $p \circ p = p$ for all $p \in B$) and antisymmetric (as $p \leq q \leq p$ means $q = p \circ q = q \circ p = p$). To see that it is transitive let $p \leq q \leq r$. Then $p \circ q = q$ and $q \circ r = r$ and so $p \circ r = p \circ (q \circ r) = (p \circ q) \circ r = q \circ r = r$ proving $p \leq r$. Let p, q and r satisfy $p \leq r \geq q$. Then $r = p \circ r = p \circ (q \circ r) = (p \circ q) \circ r$ and so $(p,q) \in N$, $p \circ q \in \{p,q\}$ i.e. p and q are comparable proving that $\leq$ is an oriented forest and that (i) holds.

By Corollary 3.12 the groupoid $<A; \cdot >$ is commutative and idempotent. By Proposition 3.4 one of the conditions (i) - (v) is satisfied. However, $\underline{B}$ is neither right nor left trivially associative and also (i) does not hold and so (v) is true proving that $<A; \cdot >$ is a semilatice. By Corollary 3.12 $< A ; + >$ is a semilattice and so $\underline{A}$ is a bisemilattice satisfying a). To show b) let $k, a, b_1, ..., b_k, c_1, ..., c_k$ be as in b) . By assumption there is a chain $d_1 < ... < d_k$ in $(B, \leq)$. Put $Z := \{ a_1, ..., a_k \}$, $a_{d_l} := 0$ for $l = 1, ..., k-1$, $a_{d_k} := a$ and $b_{d_l} := b_l$, $c_{d_l} := c_l$ for $l = 1, ..., k$. Finally let $z := d_k$. As $d_k \circ d_i \circ d_j = d_k$ for all $1 \leq i, j \leq k$ the equation (3) reduces to (6). $(C) \Rightarrow (A)$. By Corollary 3.12 the groupoid $<A; + >$ is a semilattice and $<D ; * >$ is commutative and idempotent. To verify (v) in Proposition 3.4 we must show that the pseudo-distributive law (3) holds. First note that from the first half of (C) (i) we obtain:

Fact: *Let* $p,q,r,z \in B$. *Then* $p \circ q \circ r = z$ *if and only if* p,q *and* r *form a chain whose greatest element is* z.

Next note that (6) and $\underline{A}$ semilattice imply that

$$\sum_{j=1}^{k} \left(\sum_{i=1}^{k} ab_i \right) c_j = \sum_{i=1}^{k} \left(\sum_{j=1}^{k} ac_j \right) b_i \qquad (7)$$

holds for all $k < m_{\leq}$ and $a, b_1, ..., b_k, c_1, ..., c_k \in A$.

Let Z be a finite subset of B and $z \in B$. In view of Fact we may assume that z is the greatest element of Z. Taking into account that $\leq$ is an oriented forest, we obtain that Z is a chain. If follows that $k := |Z| < m_{\leq}$. Since $<A ; + >$ is a semilattice, we may group the terms on the left side of (3) according to $p=z$, $q=z$ adnd $r = z$:

$$a_z \left(\sum_{q,r \in Z} b_q c_r \right) + \sum_{p \in Z} a_p \left(\sum_{r \in Z} b_z c_r \right) + \sum_{p \in Z} a_p \left(\sum_{q \notin Z} b_q c_z \right). \qquad (8)$$

Applying (6), (7) and (6) to the first, second and third term we transform (8) into

$$\sum_{r \in Z} \left(\sum_{q \in Z} a_z b_q \right) c_r + \sum_{r \in Z} \left(\sum_{p \in Z} a_p b_z \right) c_r + \left(\sum_{p,q \in Z} a_p b_q \right) c_z$$

which is the right side of (3). $\square$

We show that $\underline{D}$ is a lattice (with join + and meet *) only in the case of a subdirect power of a lattice; in other words, that convolutive algebras are useless for the construction of lattices.

3.14 Proposition. *The absorptive law* $f * (f + g) = f$ *holds in* $\underline{D}$ *if and only if*

(i) $\qquad a(a + b) = a$ *for all* $a, b \in A$,

(ii) $\underline{B}$ *is idempotent, and*

(iii)   *If* $N \neq P := \{ <p,p> : p \in B \}$, *then* $p \circ q = p$ *for all* $(p,q) \in N$ *and*

       $a + ab = a$ *for all* $a,b, \in A$.

Proof: Necessity. From $|a|_p * (|a|_p + |b|_p) = |a|_p$ we obtain both (i) and (ii).

Let $(p,q) \in N, p \neq q$. Suppose $p \circ q = z \neq p$. Then $|a|_p * (|a|_p + |a+b|_q)$

$= |a|_p$ at z shows $a(a+b) = 0$ contradicting (i). Thus $p' \circ q' = p'$ for all (p',q')

$\in N$. Moreover, from $|a|_p * (|a|_p + |b|_q) = |a|_p$ we get $a + ab = a^2 + ab = a$ for all

$a,b \in A$.

Sufficiency: Direct check. $\square$

3.15 Corollary. *The structure* $\underline{D}$ *is a lattice (with join* $+$ *and meet* $*$) *if and only*

if $\underline{A}$ *is a lattice with* 0, $N = P$ *and* $p \circ p = p$ *for all* $p \in B$.

Proof: Necessity. By 3.13 (or 3.14) $\underline{B}$ is idempotent. Moreover $\underline{B}$ is

commutative by 3.2, hence $N = P$ (since otherwise by 3.14 (iii) $p = p \circ q =$

$q \circ p = q$). Moreover, for $a,b \in A$ from $|a|_p + (|a|_p * |b|_p) = |a|_p$ we obtain

$a + ab = a$. Now $\underline{A}$ is a bisemilattice by 3.13 and $\underline{A}$ satisfies both absorptive

laws, hence $\underline{A}$ is a lattice.

Sufficiency. Obvious. $\square$

3.16 Remark. If $N = P := \{ (p,p) : p \in B \}$ and $p \circ p = p$ for all $p \in B$, then the

family $\underline{C}$ from 2.1 is just an order ideal in $(2^B, \subseteq)$ containing all singletons (as

$X \circ Y = X \cap Y \subseteq X$ and $\{ (x,y) \in N \cap X \times Y : x \circ y = b \} \subseteq \{b\}$) .

4. Identity and cancellation.

     We may ask under what conditions $e \in D$ is a left or right identity

(unit or neutral element) for $*$. Given $p,q \in B$ and $a \in A$ put $\delta_{pq} (a) := a$ if

$p = q$ and $\delta_{pq}(a) := 0$ otherwise. We have:

4.1 Proposition. *Let* $e \in D$. *Then* e *is a left (right) convolutive identity if and only if*

$$\sum_{xop=q} e(x)\, a = \delta_{pq}(a) \qquad \left(\sum_{pox=q} a\, e(x) = \delta_{pq}(a)\right) \qquad (9)$$

for all $p,q \in B$ and $a \in A$.

Proof: The necessity can be proved by the direct computation of $(e * \lceil a \rceil_p)(z) =$ $\lceil a \rceil_p(z)$. For the sufficiency let $f \in D$ and let e satisfy the first half of (9). Then

$$(e * f)(z) = \sum_{y \in B} \sum_{xoy=z} e(x)\, f(y) = \sum_{y \in B} \delta_{yz}\,(f(y)) = f(z)$$

for every $z \in B$. $\square$

A semiring (cf 3.9) is a *diod* if it has both 0 (which is neutral for $+$ and an annihilator for $\cdot$) and a multiplicative identity. An example of a dioid is $\underline{A} := <\mathbb{R} \cup \{\infty\}\,;\, \min, + >$ with zero ∞ and identity 0. The corresponding structure from Example 2.3 (of $n \times n$ matrices over $\underline{A}$) has applications in paths of graphs (cf Gondran & Minoux (1979) ch. 3 pp 65-80).

4.2 Corollary. *If* $\underline{A}$ *is right distributive, then* e *is a left convolutive identity if and only if for all* $p,q \in B,\, p \neq q$ *the elements*

$$\sum_{xop=p} e(x)\quad,\quad \sum_{xop=q} e(x) \qquad (10)$$

are a left identity and a left annihilator of $\underline{A}$ *, respectively.*

In particular ,the convolutive algebra is a dioid if and only if $\underline{A}$ *is a dioid (with identity* 1*),* $\underline{B}$ *is a partial semigroup and there is* $e: B \to A$ *such that*

$$\sum_{xop=p} e(x) = 1,\quad \sum_{xop=q} e(x) = 0$$

for all $p, q \in B$, $p \neq q$.

Proof: The first statement follows from (9) . Necessity for the second statement. By what has already been shown the structure $\underline{A}$ has both left and right identities, hence an identity 1 and so $A^2 \cdot A \neq \{0\} \neq A \cdot A^2$. From (10) there exists x such that $x \bullet p = p$. Then $x \circ (x \bullet p) = p$ proving that B is not trivially associative. Now from 3.9 (C) we obtain that $\underline{A}$ is a semiring (hence a dioid) and $\underline{B}$ a partial semigroup. By the already proved first part, the second sum in (9) is a left annihilator ε and so $\varepsilon = \varepsilon 1 = 0$. The proof of sufficiency is analogous. $\square$

We say that v is *a left unit of* $\underline{B}$ *if* $(v,p) \in N$ *and* $v \circ p = p$ for all $p \in B$.

4.3   Corollary. If $\underline{A}$ *has a left identity* i *and* $\underline{B}$ *has a left unit* v, *the peak function* lil_v *is a left convolutive identity.*

4.4 Corollary. Suppose that α) *for all* $p \in B$ *the equation* $x \circ p = p$ *has at most one solution* $x \in B$ *and* β) *to each* $(r,s) \in N$ *there exists* $p \in B$ *such that* $(r,p) \in N$ *and* $x \circ p = r \circ p \Leftrightarrow x = r$. *Then* $e \in D$ *is* a *left convolutive identity if and only if there is a selfmap* φ *of* B *such that* (i) *for all* $x \in$ im φ *we have* a) $e(x)$ *is a left identity of* $\underline{A}$ *and* b) $(x,y) \in N \Leftrightarrow x = \varphi(y)$, *and* (ii) $e(x)$ *is a left annihilator of* $\underline{A}$ *whenever* $x \in B \setminus$ im φ *and* $(x,y) \in N$ *for some* $y \in B$.

Proof: Necessity. Let $p \in B$. In view of (9) and α) there is a unique $\varphi(p) \in B$ such that $\varphi(p) \circ p = p$. Moreover, $e(\varphi(p))$ is a left identity of A proving a).

Suppose $r \in$ im φ and $(r,p) \in N$ where $\varphi(p) \neq r$. From α) we have that $q : = r$ $\circ p \neq p = \varphi(p) \circ p$ and from β) and (9) we obtain that $e(r)$ is a left annihilator of $\underline{A}$ in contradiction to $e(r)$ left identity. Thus b) holds. To prove (ii) suppose

$r \in B \setminus \text{im } \varphi$ and $(r,s) \in N$. By β) there is $p \in B$ such that $(r,p) \in N$ and $x \circ p$
$= r \circ p$ iff $x = r$. Setting $q := r \circ p$ we have $p \neq q$ and from (9) we obtain (ii).
Sufficiency. Direct check of (9). $\square$

4.5 Corollary. *Let* $\underline{B}$ *be such that for all* $p,q \in B$ *the equation* $x \circ p = q$ *has at most one solution* x. *Then* $e \in D$ *is a left convolutive identity if and only if there is* $\varphi : B \to B$ *satisfiying the conditions* (i) *and* (ii) *from Corollary 4.4.*

The structure $\underline{D}$ may possess left identities in other cases as well.

4.6 Example. Suppose that there are $e \in A$ and an integer $k > 0$ such that (i)
$k(ca) := ca + \ldots + ca = a$ for all $a \in A$. (ii) $|\{ x \in B : x \circ p = p \}| = k$ for all $p \in B$
and (iii) $(x,y) \in N \Rightarrow x \circ y = y$ for all $x,y \in B$. Let e be defined by $e(x) = c$
if $x \circ p = p$ for some $p \in B$ and by $e(x) = 0$ otherwise. Then e is a left
convolutive identity (the condition (i) is e.g. satisfied if $\underline{A}$ is right distributive
and $kc := c + \ldots + c$ is a left identity of $\underline{A}$).

4.7 Example. Let $B = \{0,1,2\}$ and let $\circ$ be defined by the following table

$\circ$	0	1	2
0	2	0	2
1	0	1	0
2	2	0	0

Then for each $\underline{A}$ with identity 1 and its additive inverse -1 the function e defined
by $e(0) = e(1) = 1$, $e(2) = -1$ is the identity of $\underline{D}$.

It seems that explicit conditions for the existnce of (a left) convolutive
identity vary with the structures of $\underline{A}$ and $\underline{B}$. In the direction of rings we have the
following small observation. Let e be a left convolutive identity of $\underline{D}$. For each

$p \in B$ put $\bar{p} := \{\, x \in \mathrm{supp}\ e : x \circ p = p \,\}$

4.8 Proposition. *Let* $\underline{B}$ *be associative and let* e *be a left convolutive identity of* $\underline{D}$. *Then for each* $p \in B$ *the set* $\bar{p}$ *is a finite nonempty subgroupoid of* $\underline{B}$. *Moreover* $\bar{q} \subseteq \bar{p}$ *for each* $q \in \bar{p}$ *and to each* $p \in B$ *there exists* $q \in \bar{p}$ *such that* $\bar{r} = \bar{q}$ *for all* $r \in \bar{q}$.

Proof. Let $p \in B$. From (9) for each $a \in A$

$$\sum_{x \in \bar{p}} e(x)\, a \;=\; a,$$

hence $\bar{p} \neq \phi$. Let $x, y \in \bar{p}$. From $y \circ p = p$, $x \circ p = p$ and the partial associative law it follows $(x,y) \in N$, $(x \circ y, p) \in N$ and $(x \circ y) \circ p = p$ proving $x \circ y \in \bar{p}$ and that $\bar{p}$ is a finite nonempty subgroupoid of $\underline{B}$. A similar argument shows $\bar{q} \subseteq \bar{p}$ for each $q \in \bar{p}$. Choosing $q_0 = p$ and $q_i \in \bar{q}_{i-1}$ for $i = 1,2,\ldots$ we construct $\bar{p} \supseteq \bar{q}_1 \supseteq \bar{q}_2 \supseteq \ldots$. Since $\bar{p}$ is finite, we must arrive at q_i such that $\bar{r} = \bar{q}_i$ for all $r \in q_i$. $\square$

We may ask about the existence of convolutive identity in the situation of Proposition 3.13 . For an order $(B, \leq)$ an element $m \in B$ is *minimal* if $p \leq m$ implies $p = m$. Let $M_{\leq}$ denote the set of minimal elements of $(B, \leq)$.

4.9 Proposition. *Let* $\underline{A}$ *be a bisemilattice and let* $\underline{B}$ *be a partial semilattice such that* $p \circ q \in \{p,q\}$ *for all* $(p,q) \in N$. *Then* $\underline{D}$ *has the convolutive identity if and only if* (*) $<A ; \cdot >$ *has the identity* 1, $M_{\leq} \in \underline{C}$ *and the set* $\{\, m \in M_{\leq} : m \leq b \,\}$ *is finite and nonempty for all* $b \in B$.

If (*) *holds, then the covolutive identiy* e *is determined by* $e(m) := 1$ *whenever* m *is minimal and* $e(m) := 0$ *otherwise.*

Proof: Necessity. Let e be the convolutive identity of $\underline{D}$. Let $p \in B$ be such that

$q < p$ for some $q \in B$. Then $x \circ q = p$ has a unique solution $x = p$ and from (9) we get $e(p) a = 0$ for all $a \in A$. In particular, $e(p) = e(p)e(p) = 0$. For $m \in M : = M_{\leq}$ the equation (9) gives $e(m) a = a$ for all $a \in A$, i.e. $e(m) = 1$, proving $M = \text{supp } e \in \underline{C}$. Next for $b \in B$ and $X : = \{b\} \in \underline{C}$ and $Y : = M \in \underline{C}$ the finite set from 2.1 (ii) is

$$F : = \{ (b,m) \in N : m \in M, \, b \circ m = b \}.$$

Clearly $| \{ m \in M : m \leq b \} | = |F| < \aleph_U$. Finally from (9)

$$1 = \sum_{x \circ b = b} e(x) 1 = \sum_{m \in M, \, m \leq b} 1$$

proving that $\{ m \in M : m \leq b \}$ is nonempty for all $b \in B$.

Suffiiency: Direct check of (9). $\square$

Next we apply convolution to boolean algebras. In view of 3.15 we should not consider a boolean algebra as a lattice with complementation but rather as a boolean ring (i.e. an associative and idempotent ring with identity 1 in which $a \wedge b : = ab$, $a \vee b : = a + b + ab$ and $a' : = 1 + a$ for all a,b). For $p,q \in B$ put $p \leq q$ if $(p,r) \in N$ and $p \circ r = q$ for some $r \in B$. Further put $(q] : = \{ p \in B : p \leq q \}$. We have:

4.10 Proposition. *The convolutive algebra is a boolean ring if and only if*

(i) $\underline{A}$ *is a boolean ring and*

(ii) $\underline{B}$ *is a partial semilattice and there is* $E \in \underline{C}$ *such that a)* $| E \cap (p] |$ *is finite and odd for all* $p \in P$ *and b)*

   $| \{ x \in E : (q,x) \in N, \, q \circ x = r \} |$ *is finite and even for all pairs* $q < r$.

If (i) *and* (ii) *hold, then* e *defined by setting* $e(x) : = 1$ *for* $x \in E$ *(where* 1 *is the identity of* $\underline{A}$*)* *and* $e(x) : = 0$ *otherwise is the identity of* $\underline{D}$*.*

Proof: Necessity. Let e be the identity of D. By 3.10 $\underline{A}$ is a ring. It is well known that $\underline{D}$ is commutative, hence by 3.11 (i) both $\underline{A}$ and $\underline{B}$ are commutative and idempotent. Next $\underline{B}$ is neither left nor right trivially associative and from $(e * e) * e = e = e * (e * e)$ it follows that $A^2 \cdot A \neq \{0\} \neq A \cdot A^2$. Now from 3.4 we obtain that both $\underline{A}$ and $\underline{B}$ are associative. Thus $\underline{A}$ is a boolean ring (with identity 1) and $\underline{B}$ is a partial semilattice. It is easy to see that $(B, \leq)$ is a partial order in which $p \circ q$ is the sup (least upper bound) of p and q whenever $(p,q) \in N$. From (10) we obtain

$$\sum_{x \leq p} e(x) = 1, \quad \sum_{q \circ x = r} e(x) = 0 \qquad (11)$$

for all $p,q, \in P$, $q < r$ (for $q \neq r$, $q \not< r$ the second sum in (9) is trivially 0). Put $E := \mathrm{supp}\ e$. Now $E \in \underline{C}$ and for $X := E$, $Y := \{p\} \in \underline{C}$ and $b := p$ the condition from 2.1 yields that $E \cap (p]$ is finite. We show that $e(p) := 1$ for all $p \in E$ by induction on the maximum cardinality $n(p)$ of a chain $p = x_1 > ... > x_{n(p)}$ in $E \cap (p]$. If $n(p) = 1$ (i.e. p is a minimal element of E), $e(p) = 1$ follows from the first equation of (11). Suppose $k > 1$ and $e(s) = 1$ holds for all $s \in E$ with $n(s) < k$. Let $p \in E$ satisfy $n(p) = k$. Put $F := (E \cap (p]) \setminus \{p\}$. By assumption $e(x) = 1$ for all $x \in F$ and from the first equation in (11) we obtain $e(p) + |F| 1 = 1$. It is well known and easy to prove that $1 + 1 = 0$ in $\underline{A}$. If $|F|$ were odd we would get $e(p) = 0$ and $p \notin E$. Thus $|F|$ is even and we obtain the required $e(p) = 1$. This concludes the proof by induction. Moreover, we have shown the second statement of a). Finally b) is just a translation of the second equation of (11). Sufficiency: Verification based on 3.10, 3.11 and 4.2. $\square$

4.11 Remarks 1) Let E satisfy 4.10. (ii) a). Then to each $p \in B$ we have $m \leq p$ for some minimal element of $(B, \leq)$ and E contains all minimal elements of $(B, \leq)$. (Indeed, for a minimal element p of $(B, \leq)$ the condition a) yields $\phi \neq E \cap (p] \subseteq \{p\}$ and so $p \in E$; for $q \in B$ a minimal element of the finite non-void set $E \cap (q]$ is a minimal element of $(B, \leq)$).

2) If there is a map φ from B into positive integers such that all descending chains

from p are of cardinality at most $\varphi(p)$, then by induction we can either construct E

satisfying (ii) or prove there is no such E (at each step we decide whether $p \in E$

or $p \notin E$ so that a) holds and then check b) for all $q < r = p$; clearly b) depends

also on N and not only on $\leq$ and the situation does not seem to be transparent).

3) If $(B, \leq)$ has a least element ℓ , then $E = \{\ell\}$.

4.15 Example. Let B be the set of positive integers and let $x \circ y$ be the least

common multiple of x and y . Let $\underline{A} = < \{0,1\} ; + , \cdot , 1 >$ be the two element

boolean ring (where $0 + 1 = 1 + 0 = 1, 0 + 0 = 1 + 1 = 0$). Let $\underline{C}$ be an order

ideal in $(2^B, \subseteq)$ whose union is B. Finally let $e : B \to \{0,1\}$ be defined by

$e(1) : = 1$ and $e(x) : = 0$ otherwise. Then $\underline{D} = < D ; + , *, \underline{0}, e >$ is a boolean

ring.

We conclude with remarks on cancellation properties for which we only

have very partial results. We say that a groupoid $<X ; \cdot >$ with 0 is *left*

cancellative if $xy = xz$ implies $y = z$ for all $x \in X^* : = X \setminus \{0\}$ and $y,z \in X$.

4.16 Proposition. *If the convolutive algebra is left cancellative, then:*

(i) $\underline{B}$ *is a full groupoid such that for all* $p,q \in B$ *the equation* $p \circ x = q$ *has*

        at most one solution $x \in B$,

(ii)   *$<A ; \cdot >$ is left cancellative,*

(iii)   *if P,Q,R are finite non empty subsets of B such that $Q \cap R = \phi$, then to*

        each $c \in A^2 \setminus \{0\}$ there exists $z \in B$ such that $z = p \circ q$ has exactly ν

        solutions $(p,q) \in P \times Q$ and $z = p \circ r$ has exactly μ solutions $(p,r) \in$

        $P \times R$ where $\nu c \neq \mu c$,

(iv)   *the map $x \to a_1 x + ... + a_k x$ is an injective selfmap of A whenever*

        $a_1,...,a_k \in A^*$ and $p_1 \circ q = ... = p_k \circ q$ for some pairwise distinct

$p_1,\ldots,p_k \in B$ *and some* $q \in B,$

(v) *if* $\underline{A}$ *is right distributive, then either* (a) *for all* $q, r \in B$ *the equation* $x \circ q = r$ *has at most one solution* $x \in B$ *or* (b) A^* *is a subgroupoid of* $\langle A ; + \rangle.$

Proof: Fix $a \in A^* := A \setminus \{0\}.$

(i) Suppose $(p,q) \notin N.$ Then $\underline{0} = \lceil a \rceil_p * \lceil a \rceil_q = \lceil a \rceil_p * \underline{0}$ and this contradiction shows $N = B^2$ i.e. that B is a (full) groupoid. Suppose $p \circ x = p \circ x'.$ From $\lceil a \rceil_p * \lceil a \rceil_x = \lceil a \rceil_p * \lceil a \rceil_{x'}$ we get $x = x'.$

(ii) Fix $p \in B.$ If $a \in A^*$ and $ay = ay'$ for some $y, y' \in A$ then $\lceil a \rceil_p * \lceil y \rceil_p = \lceil a \rceil_p * \lceil y' \rceil_p$ shows $y = y'$ i.e. $\underline{A}$ is left cancellative.

(iii) Let P, Q, R and c be as in the hypothesis. Then $c = ab$ for some $a, b \in A^*.$ Let $f(x) := a$ for $x \in P$ and $f(x) := 0$ otherwise, $g(x) := b$ for $x \in Q$ and $g(x) := 0$ otherwise and $h(x) := b$ for $x \in R,$ $h(x) := 0$ otherwise. Clearly

$g \neq h$ and so $f * g \neq f * h$ proving the existence of $z \in B$ with the required properties.

(iv) Let $a_1,\ldots,a_k$, $p_1,\ldots,p_k$ and q be as in the hypothesis. Put $u := p_1 \circ q$ and $\varphi(x) := a_1 x + \ldots + a_k x$ for all $x \in A.$ Suppose $\varphi(t) = \varphi(t')$ for some $t, t' \in A.$ Put $f := \lceil a_1 \rceil_{p_1} + \ldots + \lceil a_k \rceil_{p_k}$, $h := f * \lceil t \rceil_q$ and $g := f * \lceil t' \rceil_q.$ To prove $t = t'$ it suffices to show $h = g.$ Indeed, for $z \in B$ we have $h(z) \neq 0 \Leftrightarrow z = u,$ $h(u) = \varphi(t) \Leftrightarrow z = u,$ $g(u) = \varphi(t') \Leftrightarrow g(z) \neq 0.$

(v) Let $\underline{A}$ be right distributive and suppose there are $p_1, p_2, q \in B,$ $p_1 \neq p_2$ such that $p_1 \circ q = p_2 \circ q.$ Let $a_1, a_2 \in A^*.$ According to (iv) the map $\varphi(x) = (a_1 + a_2)x$ is injective and so $a_1 + a_2 \neq 0$ i.e. $a_1 + a_2 \in A^*.$ $\square$

We say that a groupoid with 0 is *cancellative* if it is both left and right cancellative.

4.17 Corollary. *If the convolution is cancellative, then* (i) $\underline{B}$ *is a full infinite groupoid in which for all* $p, q \in B$ *we have* a) *both equations* $p \circ x = q$ *and* $x \circ p = q$ *have at most one solution and* b) *for* $p \neq q$ *and each* $P \subseteq B$ *finite there exist* $r, r' \in B$ *such that exactly one of the equations* $p \circ x = r$, $q \circ x = r$ ($x \circ p = r'$, $x \circ q = r'$) *has a solution in* P , *and* (ii) $\underline{A}$ *is cancellative.*

Proof: We only prove that B is infinite since all the other statements follow from 4.16 (e.g for the second statement of b) choose $Q := \{q\}$, $R := \{p\}$ and $z = r'$). Suppose B is finite. Then $\underline{B}$ is a quasigroup in contradiction to b) (with $P := B$). $\square$

4.18 Example. *If* $(x,y) \rightarrow x \circ y$ *is an injective map from* B^2 *into* B *and* $\underline{A}$ *is cancellative, then the convolution is cancellative.*

Bibliography

BERGERON, F., (1986). Une systématique de la combinatoire énumérative. Thèse Ph.D. Université de Montréal.

CASHWELL, E.D. & EVERETT, C.J., (1959). The ring of number-theoretic functions. Pacific J. Math. 9, 975-85.

CONTENT, M., (1977). Les catégories de Möbius. Mémoire de maîtrise, UQAM, Montréal, 99 pp.

CONTENT, M. & LEMAY, F. & LEROUX, P., (1980). Catégories de Möbius. J. Combinatorial Theory Ser A, 28, No 2, 169-90.

DEZA, M.-M. & ROSENBERG, I.G., (1986). General convolutions motivated by designs. Acta Universitatis Carolinae - Math. et Phys, Vol. 27, no 2, 49-65.

FINCH, P.D., (1970). On the Möbius function of a non-singular binary relation. Bull. Austral. Math., Soc. 3, 155-62.

GONDRAN, M. & MINOUX, M., (1979). Graphes et algorithmes. Eyerolles.

GRAVER, J.E. & JURKAT, W.B., (1972). Algebra structures for general designs. Journal of Algebra, 23, no 3, 574-89.

GRAVER, J.E. & JURKAT, W.B., (1973). The module structure of integral designs. J. Combinatorial Theory (A), 15, 75-90.

HEBISCH, V. & WEINERT, H.J., (1985). Euclidean seminearrings and nearrings. Proceedings Internat. Conference on Near-rings and Near-fields, (Tübingen 1985), North Holland.

JOYAL , A., (1981). Une théorie combinatoire des séries formelles. Advances in Math., 42, no 1, 1-82.

LABELLE, J., (1983). Applications diverses de la théorie combinatoire des espèces de structures, Ann. Sci. Math du Québec, 7 , no 1, 59-94.

LAUGWITZ, D., (1968). Eine nicht archimedische Erweiterung angeordneter Körper. Math. Nachr, 37, 225-36.

LEMAY, F., (1977). Catégories de Möbius, algèbres d'incidence et fonctorialités. Mémoire de maîtrise, UQUAM Montréal, 70 pp.

LEROUX, P., (1975). Les catégories d Möbius, Cahiers de topologie et géométrie différentielle, Vol. 16, pp. 280-282.

LUNC, A.G., (1950). The application of boolean matrix algebra to the analysis and synthesis of relay-contact networks (Russian). Dokl. Akad Nauk USSR, 70, 421-3.

NIVEN, I., (1969). Formal power series. Amer. Math. Monthly, $\underline{76}$, 871-89.

ROMANOWSKA, A., (1983). Building bisemilattices from lattices and semilattices. <u>In</u> Contributions to General Algebra 2 (Proceedings Klagenfurt Conference June 1982) Hölder-Pichler-Tempsky Wien & B.G. Teubner Stuttqart, pp 343-57.

ROMANOWSKA, A., (1982). Algebras of functions from partially ordered sets into distributive lattices. Proceedings Conf. Universal Algebra and Lattice Theory (Puebla, Mexico, Freese R.S. & Garcia O.C. eds). Lecture Notes in Mathematics 1004, Springer-Verlag, Berlin Heidelberg New York Tokyo.

ROSENBERG, I.G., (1977). A generalization of semigroup rings. Preprint CRM-732, Université de Montréal, 22 pp.

SALOMAA, A. & SOITTOLA, M., (1978). Automata-theoretic aspects of formal power series. Texts & Monogr. in Comput. Sc. (Bauer, F.L. & Gries, D. eds) Springer Verlag New York.

WEINERT, J.H., (1984). On 0-simple semirings, semigroup semiring and two kinds of division semirings. Semigroup Forum, $\underline{28}$, 313-33.

QUASIGROUPS, ASSOCIATION SCHEMES, AND LAPLACE OPERATORS ON
ALMOST PERIODIC FUNCTIONS

J.D.H. Smith
Iowa State University, Ames, Iowa   50011

Abstract. The association scheme determined by a finite non-empty quasigroup furnishes generalized Laplace operators Δ_i on the space of almost periodic functions on the free group that is the universal multiplication group of the quasigroup. An existence theorem is proved for solutions of the equation $\Delta_i u = 0$ on the closed convex hull of the set of twisted translates of a given almost periodic function f. This theorem generalizes the classical result on the existence of von Neumann means of almost periodic functions.

INTRODUCTION

There is a well-known and intimate connection between the
character theory and the ordinary representation theory of a finite
group: characters are traces of matrix representations, and determine
the representations up to equivalence. For a finite non-empty
quasigroup Q, the connections are much more obscure. The character
theory is that of the association scheme $\left(Q{\times}Q;\ C_1,\ldots,C_s\right)$ determined
by the permutation representation of the combinatorial multiplication
group on the quasigroup [1, pp. 181-2] [3] [6]. The representation
theory is that of free groups determined by the quasigroup -- the
universal multiplication group and point stabilizers within it [6].
Representations are classified by almost periodic functions on these
free groups. The character theory furnishes generalized Laplace
operators $\Delta_1,\ldots,\Delta_s$ acting on the almost periodic functions. Thus to
understand the connections between the character theory and the
representation theory, one needs to study the behavior of these Laplace
operators on almost periodic functions.

One of the major classical theorems about almost periodic
functions is the existence of the von Neumann mean of an almost periodic
function. For almost periodic functions on the universal multiplication
group of a quasigroup, this theorem was interpreted in [6, 645] as an

existence and uniqueness theorem for solutions of Laplace's equation
$\Delta_1 u = 0$ on the closed convex hull of the set of translates of an almost
periodic function. This paper presents a more general existence result
(Theorem 4.5) for solutions of $\Delta_i u = 0$ on the closed convex hull of
the set of certain twisted translates (4.1) of an almost periodic
function. An example (Example 4.6) shows that in general the solutions
are not unique. Indeed, Problem 4.7 raises the question of describing
the shape of the solution sets in general. For periodic functions this
is a purely combinatorial problem. An analytical problem (Problem 4.4)
also arises in trying to characterize almost periodic functions in terms
of their sets of twisted translates.

The main Theorem 4.5 is proved in Section 5. Sections 2 and
3 provide a quick sketch of the background required, particularly for
readers acquainted with association schemes as described in [1, Chapter
2] or [6, Chapter 5]. The analytical proof in Section 5 has been
written out in considerable detail to make it more readily understood by
readers having only an elementary acquaintance with topological groups.

QUASIGROUPS AND ALMOST PERIODIC FUNCTIONS

A _quasigroup_ is a set Q equipped with three binary
operations • (called _multiplication_), /, and \, such that the
identities

$$(2.1) \quad \begin{cases} (x/y)\cdot y = x \; ; \\[4pt] (x\cdot y)/y = x \; ; \\[4pt] x\cdot(x\backslash y) = y \; ; \\[4pt] x\backslash(x\cdot y) = y \end{cases}$$

are satisfied. These identities imply that for each element q of
Q, the mappings

$$(2.2) \qquad R(q) : Q \to Q \; ; \; x \mapsto x\cdot q$$

and

$$(2.3) \qquad L(q) : Q \to Q \; ; \; x \mapsto q\cdot x$$

are permutations of the set Q. The (_combinatorial_) _multiplication
group_ G of Q is the subgroup of the permutation group of Q

generated by $\{R(q),L(q) \mid q \in Q\}$. If Q is finite and non-empty, then
the permutation representation of G on Q determines a (commutative)
association scheme: the orbits $C_1 = \hat{G}$, $C_2,\ldots,C_s$ of G acting
diagonally on $Q \times Q$ partition $Q \times Q$, and the **C**-linear span of the
adjacency matrices $\underline{C}_1 = \underline{I}$, $\underline{C}_2,\ldots,\underline{C}_s$ of these orbits (considered as
relations on Q) is a commutative algebra $V(G,Q)$, the <u>Bose-Mesner</u>
<u>algebra</u> or <u>centraliser ring</u> of G on Q [1] [3] [6].

  Quasigroups as defined by (2.1) form a variety in the sense
of universal algebra. Let I be the free quasigroup on 1 generator,
and let $\tilde{Q}$ be the coproduct of Q with I in the category $\underline{\underline{Q}}$ of
quasigroup homomorphisms. Define mappings

(2.4) $\tilde{R}(q) : \tilde{Q} \rightarrow \tilde{Q} ; x \mapsto x \cdot q$

and

(2.5) $\tilde{L}(q) : \tilde{Q} \rightarrow \tilde{Q} ; x \mapsto q \cdot x$

for each q in Q. In the representation theory of the quasigroup Q
[6], one studies the subgroup $\tilde{G}$ of the multiplication group of $\tilde{Q}$
generated by $\{\tilde{R}(q),\tilde{L}(q) \mid q \in Q\}$. This group $\tilde{G}$ is called the
<u>universal multiplication group</u> of Q. It is free on
$\{\tilde{R}(q),\tilde{L}(q) \mid q \in Q\}$ [6, 238], and has the combinatorial multiplication
group G as its image under the homomorphism $\tilde{G} \rightarrow G$ defined by
$\tilde{R}(q) \mapsto R(q)$ and $\tilde{L}(Q) \mapsto L(q)$ for all q in Q. Under the discrete
topology, the infinite group $\tilde{G}$ is locally compact but not compact.
However, the forgetful functor from the category of compact topological
groups to the category of all topological groups has a left adjoint
called <u>Bohr compactification.</u> Let $\tilde{G}^b$ denote the Bohr compactification
of $\tilde{G}$. Since the free group $\tilde{G}$ is residually finite, the universal
homomorphism $\alpha : \tilde{G} \rightarrow \tilde{G}^b$ embeds $\tilde{G}$ as a dense subgroup of $\tilde{G}^b$
([2, 16.4.1], cf. [6, Ch. 6]). The restrictions to $\tilde{G}$ of continuous
functions $\tilde{G}^b \rightarrow C$ are called <u>almost periodic functions</u> on $\tilde{G}$.

  Almost periodic functions admit an intrinsic characterization
independent of the Bohr compactification. Let $B(\tilde{G})$ denote the set of
all bounded complex-valued functions on $\tilde{G}$. This set $B(\tilde{G})$ becomes a
normed space under the <u>uniform norm</u> $\|f\| = \sup\{|f(x)| \mid x \in \tilde{G}\}$. For the
element y of $\tilde{G}$, define the <u>(left) translation operator</u> T^y on
$B(\tilde{G})$ by

(2.6) $T^y f(x) = f(y^{-1}x)$

for each x in $\tilde{G}$. (Note that $y \mapsto T^y$ furnishes a representation of $\tilde{G}$.)

PROPOSITION 2.7. A complex-valued function f on $\tilde{G}$ is almost periodic if and only if it lies in $B(\tilde{G})$, and the set

(2.8) $\{T^y f \mid y \in \tilde{G}\}$

of all left translates of f has a compact closure in the uniform norm on $B(\tilde{G})$.

<u>Proof</u>. See [2, 16.2.1] or [4, §41].

The inspiration for this paper is the following result of von Neumann [5] (originally formulated, of course, for arbitrary groups $\tilde{G}$).

THEOREM 2.9. The closed convex hull of the set $\{T^y f \mid y \in \tilde{G}\}$ of translates of an almost periodic function f on $\tilde{G}$ contains a unique constant function.

<u>Proof</u>. See [4, 41D], [5].

The constant in Theorem 2.9 turns out to be the Haar integral $\int f^b(x)dx$ over the compact group $\tilde{G}^b$ of the unique continuous function f^b on $\tilde{G}^b$ that restricts to the almost periodic function f on the dense subgroup $\tilde{G}$ [4, 41D]. The constant is called the <u>von Neumann mean</u> M(f) of the function f.

ASSOCIATION SCHEMES AND LAPLACE OPERATORS

Various spaces of functions are associated with the finite non-empty quasigroup Q, and the association scheme determined by Q furnishes so-called "generalized Laplace operators" on these spaces. Recall that, along with its basis $\{\underline{C}_1 = \underline{I}, \underline{C}_2, \ldots, \underline{C}_s\}$ of adjacency matrices of G-orbits on $Q \times Q$, the commutative C-algebra V(G,Q) also has a basis $\{\underline{E}_1 = |Q|^{-1}\underline{J}, \underline{E}_2, \ldots \underline{E}_s\}$ of orthogonal idempotents [1]

[6]. The bases are related by

$$(3.1) \qquad \underline{C}_i = \sum_{i=1}^{s} \xi_{ij} \underline{E}_j$$

and

$$(3.2) \qquad \underline{E}_i = \sum_{j=1}^{s} \eta_{ij} \underline{C}_j \, ,$$

the $(s \times s)$-matrices (ξ_{ij}) and (η_{ij}) being called the __first__ and __second__ __eigenmatrices__ respectively. For $i = 1, \ldots, s$, a function $H_i : Q \times Q \to \mathbf{C}$ is defined by setting $H_i(x,y) = \eta_{ij}$ for (x,y) in the orbit C_j. The first of the function spaces is the space $L^1(Q)$ of all functions $f : Q \to \mathbf{C}$. Then for each $i = 1, \ldots, s$, the __generalized__ __Laplace operator__ Δ_i on $L^1(Q)$ is defined by

$$(3.3) \qquad \Delta_i f(q) = f(q) - \sum_{r \in Q} H_i(q,r) f(r)$$

for q in Q. For $i = 1$, the operator Δ_i is just called the __Laplace operator__ Δ_1. Since $|Q| \underline{E}_1 = \underline{J} = \sum_{i=1}^{s} \underline{C}_i$, one has $\eta_{1j} = |Q|^{-1}$ for all $j = 1, \ldots, s$, so that

$$(3.4) \qquad \Delta_1 f(q) = f(q) - \frac{1}{|Q|} \sum_{r \in Q} f(r) \, .$$

Thus the only solutions u of "Laplace's equation" $\Delta_1 u = 0$ in the space $L^1(Q)$ are the constant functions.

The next function space to be studied is the space $P(\widetilde{G})$ of "periodic functions" on $\widetilde{G}$. A function $f : \widetilde{G} \to \mathbf{C}$ is said to be __periodic__ if it satisfies the equivalent conditions of

LEMMA 3.5 [6, 621]. For $f : \widetilde{G} \to \mathbf{C}$, the following conditions are equivalent:

(a) there is a subgroup H of finite index in $\widetilde{G}$ such that f is constant on right cosets of H;

(b) there is a subgroup H' of finite index in $\widetilde{G}$ such that f is constant on left cosets of H';

(c) there is a subgroup H'' of finite index in $\widetilde{G}$ such that f is constant on double cosets of H'';

(d) there is a normal subgroup K of finite index in $\tilde{G}$ such that f is constant on the cosets of K.

The groups H, H', H'', and K of Lemma 3.5 are called _periods_ of f. More specifically, H is a <u>left period</u>, H' is a <u>right period</u>, H'' is a <u>bilateral period</u>, and K is a <u>normal period</u>. Note that functions f: G → **C** correspond naturally to periodic functions $\tilde{f}$: $\tilde{G}$ → **C** having the kernel of the epimorphism $\tilde{G}$ → G as their normal period. A convolution * on the space $P(\tilde{G})$ is given by the following proposition.

PROPOSITION 3.6 [6, 625]. If f has bilateral period H, and g has left period K, then

$$(3.7) \qquad f*g(x) = \frac{1}{|T|} \sum_{t \in T} f(xt^{-1})g(t)$$

(for a right transversal T to $H \cap K$ in $\tilde{G}$) defines a function f*g in $P(\tilde{G})$ with left period H. The function f*g is well-defined independently of the choice of H, K, and T.

REMARK 3.8. In [1, II(11.16)], a convolution (denoted × here to distinguish it from that of (3.7)) is defined on the space of all functions from G to **C** by $(f \times g)(x) = \sum_{y \in G} f(xy^{-1})g(y)$. Under the correspondence with periodic functions on $\tilde{G}$, one then has that $\widetilde{f \times g} = |G| \, \tilde{f} * \tilde{g}$.

From now on, fix an element e of Q. Let G_e be the stabilizer of e in G, and let $\tilde{G}_e$ be the preimage of G_e in the epimorphism $\tilde{G}$ → G. A function f in $L^1(Q)$ determines a periodic function

$$(3.9) \qquad f^{\#} : \tilde{G} \to \mathbf{C} \; ; \; x \longmapsto f(ex)$$

on $\tilde{G}$ with $\tilde{G}_e$ as left period. The correspondence $f \longmapsto f^{\#}$ embeds $L^1(Q)$ in $P(\tilde{G})$. The generalized Laplace operators on $L^1(Q)$ extend to corresponding operators on $P(\tilde{G})$. For i = 1,...,s, define

$$(3.10) \qquad \mu_i : \tilde{G} \to \mathbf{C} \; ; \; x \to |Q| \cdot \bar{H}_i(e, ex) \; .$$

One may think of these μ_i as "signed measures" on the free group $\widetilde{G}$. Each μ_i clearly has $\widetilde{G}_e$ as a left period. It also has $\widetilde{G}_e$ as a right period, since for α in $\widetilde{G}_e$ one has

$$|Q|^{-1} \mu_i(x\alpha) = \bar{H}_i(e,ex\alpha) = \bar{H}_i(e\alpha^{-1},ex) = \bar{H}_i(e,ex) = |Q|^{-1} \mu_i(x).$$

PROPOSITION 3.11. For $i = 1,\ldots,s$, the periodic function μ_i is an idempotent under the convolution $*$ on $P(\widetilde{G})$.

Proof. Apply [1, §II.11] with $H = G_e$. For the function $\omega_i : G \to \mathbf{C}$ with $\omega_i(x) = \xi_{ji}/n_j$ for (e,ex) in C_j ([1, Th. II.11.10] with the notation $n_j|Q| = |C_j|$ of [3], [6]), the function $(\mathrm{tr}\ \underline{E}_i)|G|^{-1} \omega_i$ is idempotent under the convolution $\times$ of Remark 3.8 [1, Cor. II.11.7(i)]. Thus $(\mathrm{tr}\ \underline{E}_i)\widetilde{\omega}_i$ is idempotent under the convolution $*$ on $P(\widetilde{G})$. But for x in $\widetilde{G}$ with (e,ex) in C_j, one has $(\mathrm{tr}\ \underline{E}_i)\widetilde{\omega}_i(x) = (\mathrm{tr}\ \underline{E}_i)\xi_{ji}/n_j = |Q|\bar{n}_{ij} = \mu_i(x)$, the middle equality just being $(\mathrm{tr}\ \underline{E}_i)^{1/2}$ times the equality of [3, Def. 3.3] or [6, 536].

Using μ_i, the generalized Laplace operator Δ_i may be defined on $P(\widetilde{G})$ by

$$(3.12) \qquad \Delta_i f = f - \mu_i * f .$$

This extends the definition of the generalized Laplace operators on $L^1(Q)$, since $(\Delta_i f)^{\#} = \Delta_i(f^{\#})$ for f in $L^1(Q)$ [6, 626].

The last of the principal function spaces to be considered is the space $AP(\widetilde{G})$ of almost periodic functions on $\widetilde{G}$. Since a periodic function only has finitely many translates, Proposition 2.7 shows that $P(\widetilde{G})$ is a subspace of $AP(\widetilde{G})$. Let f be a periodic function on $\widetilde{G}$ with left period H, and let T be a right transversal to H in $\widetilde{G}$. Then the von Neumann mean of the almost periodic function f is

$$\int f^b(x)dx = \frac{1}{|T|} \sum_{t \in T} f(t) \quad [6,642].$$ Now the space $L^1(\widetilde{G}^b)$ of functions $g : \widetilde{G}^b \to \mathbf{C}$ with $\int |g(x)|dx$ defined [4, §17] has a convolution given by

$$f*g(x) = \int f(xt)g(t^{-1})dt = \int f(xt^{-1})g(t)dt \quad [4, 31A].$$ Then for periodic

functions f, g on $\tilde{G}$, one has $(f*g)^b = f^b*g^b$. Thus the generalized
Laplace operators (3.12) on $P(\tilde{G})$ may be extended to $L^1(\tilde{G}^b)$, and
hence by restriction to $AP(\tilde{G})$.

DEFINITION 3.13 [6,644]. For $i = 1,\ldots,s$, the <u>generalized Laplace</u>
<u>operator</u> Δ_i on $L^1(\tilde{G}^b)$ (and on $AP(\tilde{G})$) is defined by

$$\Delta_i f = f - \mu_i^b * f \ .$$

In particular, Δ_1 is called the <u>Laplace operator</u>. Since the
convolution $*$ in $L^1(G^b)$ is bilinear, the generalized Laplace
operators Δ_i on $L^1(\tilde{G}^b)$ and $AP(\tilde{G})$ are linear. Note, too, that
convolution with μ_1 is just Haar integration. Theorem 2.9 may thus be
given the following reformulation, in preparation for the generalization
of the existence statement that forms the main subject of this paper.

THEOREM 3.14 [6, 645]. There exists a unique solution u to Laplace's
equation $\Delta_1 u = 0$ on the closed convex hull of the set $\{T^y f \mid y \in \tilde{G}\}$
of translates of any given almost periodic function f on $\tilde{G}$.

TWISTED TRANSLATION OPERATORS

For $i = 1,\ldots,s$, and for an element y of $\tilde{G}$, define
the i-th <u>twisted translation operator</u> T_i^y on $B(\tilde{G})$ by

$$(4.1) \qquad T_i^y f(x) = \frac{\mu_i(y)}{\mu_i(1)} \, f(y^{-1}x) \ .$$

Taking the trace of (3.2) gives $0 \neq \operatorname{tr} \underline{E}_i = \eta_{i1}|Q| = \overline{\mu_i(1)}$, so that
(4.1) is always defined. The first twisted translation operator T_1^y is
just the original translation operator T^y of (2.6). In general,
however, $y \mapsto T_i^y$ need not give a representation of $\tilde{G}$. The subset
$Z_i = \{\mu_i(y)/\mu_i(1) \mid y \in \tilde{G}\}$ of $\mathbf{C}$ is finite, and contains
$1 = \mu_i(1)/\mu_i(1)$. Set $M_i = \max\{|z| \mid z \in Z_i\}$, so that $M_i \geqslant 1$. The
number M_i is called the i-th <u>modulus</u>. One direction of Proposition
2.7 has the following generalization.

PROPOSITION 4.2. For each $1 \leqslant i \leqslant s$, the set

$$(4.3) \qquad T_i f = \{T_i^y f \mid y \in \tilde{G}\}$$

of i-th twisted translates of an almost periodic function f on $\tilde{G}$
has a compact closure in the uniform norm on $B(\tilde{G})$.

<u>Proof</u>. Since f is almost periodic, Proposition 2.7 shows that for
given $\varepsilon > 0$, there is a finite subset $\{f_1,\ldots,f_r\}$ of $B(\tilde{G})$ such
that $T_i f$ is contained in the union $B_{\varepsilon/M_i}(f_1) \cup \ldots \cup B_{\varepsilon/M_i}(f_r)$ of
balls of radius ε/M_i centered on the f_j. Consider a given i-th
twisted translate $T_i^y f$ of f. Suppose $\|T_i^y f - f_j\| < \varepsilon/M_i$. Then
$\|T_i^y f - \mu_i(y)f_j/\mu_i(1)\| = |\mu_i(y)/\mu_i(1)| \cdot \|T_i^y f - f_j\| < \varepsilon$. Thus $T_i f$ is
contained in the union of the finite set $\{B_\varepsilon(zf_j) \mid z \in Z_i,\ 1 \leqslant j \leqslant r\}$
of balls of radius ε. In other words, the set $T_i f$ is totally
bounded, and hence has a compact closure (cf. [4, 41A]).

PROBLEM 4.4. Is the converse of Proposition 4.2 false for those i for
which μ_i may take the value zero?

The generalization of the existence statement of Theorem 3.14
that forms the main topic of this paper may now be stated.

THEOREM 4.5. For $1 \leqslant i \leqslant s$, the equation $\Delta_i u = 0$ has a solution on
the closed convex hull of the set $T_i f$ of twisted translates of any
given almost periodic function f on $\tilde{G}$.

Theorem 4.5 will be proved in the next section. The rest of
this section is devoted to an example showing that a solution u of
$\Delta_i u = 0$ on the closed convex hull of $T_i f$ need not be unique if
$i > 1$.

EXAMPLE 4.6. Let Q be the set of fourth roots of unity in **C**, with
quasigroup multiplication $\circ$ given by $x \circ y = \bar{x}y$ [3, §4] [6, 537]. The
orbits C_1, C_2, C_3 of $G = D_4$ on $Q \times Q$ are the relations of
identity, diametrical opposition, and adjacency on the unit circle,
respectively. Choose the second row of the second eigenmatrix to be
$\frac{1}{4}(1,1,-1)$. Taking the fixed element e of Q to be 1, $L^1(Q)$
embeds into $AP(\tilde{G})$ via $f \mapsto f^{\#}$ with $f^{\#}$ as in (3.9). For this

example, it is sufficient to work in $L^1(Q)$. Given q in Q, define

$\delta_q : Q \to \mathbf{C}$ by $\delta_q(q) = 1$ and $\delta_q(Q-\{q\}) = \{0\}$. Let $f = \delta_1$. Then

$T_2 f = \{\delta_1, -\delta_i, \delta_{-1}, -\delta_{-i}\}$. The closed convex hull of this set of trans-

lates is a geometric 3-simplex. For g in $L^1(Q)$, the effect $\Delta_2 g$

of the Laplace operator Δ_2 is defined by $\Delta_2 g(y) = \frac{3}{4} g(y) - \frac{1}{4} g(-y) +$

$\frac{1}{4} g(iy) + \frac{1}{4} g(-iy)$ for y in Q. Thus three solutions u of

$\Delta_2 u = 0$ on the closed convex hull of the set of translates $T_2 f$ are

$\frac{1}{4} \delta_1 - \frac{3}{4} \delta_i$, $\frac{1}{4} \delta_1 - \frac{3}{4} \delta_{-i}$, and $\frac{1}{4} \delta_1 + \frac{3}{4} \delta_{-1}$. The full set of

solutions is the geometric 2-simplex spanned by these three solutions.

PROBLEM 4.7.   In the context of Theorem 4.5, find a general method to
determine the shape of the full set of solutions u of $\Delta_i u = 0$ on the
closed convex hull of $T_i f$. (For f in $L^1(Q)$ or $P(\tilde{G})$, this is a
purely combinatorial problem.)

PROOF OF THE THEOREM

This section is devoted to the proof of the main Theorem 4.5.
Fix i in $\{1,\ldots,s\}$, and let f be an almost periodic function on
$\tilde{G}$. If $f = 0$, then $u = 0 \in \{0\} = T_i f$ solves $\Delta_i u = 0$. If $f \neq 0$,
say $f(x^{-1}) \neq 0$, then $T_i T^x f = T_i f$ and $T^x f(1) = f(x^{-1}) \neq 0$, so
without loss of generality one may assume $f(1) \neq 0$. For a subset X
of $\tilde{G}$, let $\chi_X : \tilde{G} \to \mathbf{C}$ denote the characteristic function with
$\chi_X(X) = \{1\}$ and $\chi_X(\tilde{G}-X) = \{0\}$. Take $|Q| = n$, and let
$\{H = H_1,\ldots,H_n\}$ be the set of right cosets of $H = \tilde{G}_e$. Then
$1 = \sum_{j=1}^{n} \chi_{H_j}$ in $B(\tilde{G})$. The χ_{H_j} are periodic (having the subgroup H
of index n as left period), and so are almost periodic on $\tilde{G}$. Under
the ring operations induced from the ring operations on $\mathbf{C}$, the set of
almost periodic functions $\tilde{G} \to \mathbf{C}$ forms a ring [4, 41A]. Set
$f_j = \chi_{H_j} f$, so that

$$(5.1) \qquad f = f_1 + \ldots + f_n$$

for almost periodic f_j vanishing off the coset H_j. Suppose that f_j is non-zero, say $f_j(h_j) \neq 0$ for $h_j \in H_j$. In particular, take $h_1 = 1$. Define a function $\phi_j : \widetilde{G} \to \mathbf{C}$ by $\phi_j(x) = f_j(xh_j)$. Then ϕ_j is almost periodic [2, 16.2.1] [6, 641], and $\phi_j(1) = f_j(h_j) \neq 0$. Let $Z = \{f_j(h_j) \mid f_j \neq 0\}$, a finite non-empty subset of $\mathbf{C} - \{0\}$. Take $m = \min\{|z| \mid z \in Z\}$. The method of proof being used in this section (based on the approach of [4, 41D]) is to show that, given ε with $0 < \varepsilon < mnM_i$, there is a finite convex combination of translates of f differing by less than ε from a fixed solution u of $\Delta_i u = 0$ in the uniform norm on $B(\widetilde{G})$. Now the almost periodic function ϕ_j on $\widetilde{G}$ is the restriction of a unique continuous function ϕ_j^b on $\widetilde{G}^b$. Thus there is a neighborhood V_j of the identity in $\widetilde{G}^b$ such that $V_j = V_j^{-1}$ and

$$(5.2) \qquad xy^{-1} \in V_j \;\Rightarrow\; |\phi_j^b(x) - \phi_j^b(y)| < \varepsilon/nM_i \;.$$

If $f_j = 0$, take V_j to be V_1. Then set $V = \bigcap_{j=1}^{n} V_j$. Note $V = V^{-1}$.

LEMMA 5.3. (i) The neighborhood V is contained in the closure $\bar{H}$ of $\widetilde{G}_e$. Indeed, $V \cap \widetilde{G} \subseteq \widetilde{G}_e$.

(ii) For xy^{-1} in V, one has $|f^b(x) - f^b(y)| < \varepsilon/M_i$.

(iii) The function μ_i^b is constant on all translates aV of V with a in $\widetilde{G}$.

<u>Proof.</u> (i) For v in $V \subseteq V_1$, condition (5.2) gives
$$|f_1^b(v) - f_1(1)| = |\phi_1^b(v) - \phi_1^b(1)| < \varepsilon/nM_i < m < |f_1(1)|, \quad \text{so}$$
$f_1^b(v) \neq 0$. Since f_1 vanishes off $\widetilde{G}_e$, the point v must lie in the closure of $\widetilde{G}_e$. If v also lies in $\widetilde{G}$, then $f_1(v) \neq 0$ gives $v \in \widetilde{G}_e$.

(ii) Suppose $xy^{-1} \in V$. If $f_j = 0$, the inequality
$$(5.4) \qquad |f_j^b(x) - f_j^b(y)| < \varepsilon/nM_i$$

is automatic. Otherwise, write $x = \xi h_j$ and $y = \eta h_j$. Then

$$|f_j^b(x) - f_j^b(y)| = |f_j^b(\xi h_j) - f_j^b(\eta h_j)| = |\phi_j(\xi) - \phi_j(\eta)| < \varepsilon/nM_i \text{ by}$$

(5.2), since $\xi\eta^{-1} = (xh_j^{-1})(yh_j^{-1})^{-1} = xy^{-1} \in V \subseteq V_j$. From (5.1) one

obtains $f^b = \sum_{j=1}^{n} f_j^b$. Then $|f^b(x) - f^b(y)| = |\sum_{j=1}^{n} f_j^b(x) - \sum_{j=1}^{n} f_j^b(y)| =$

$$|\sum_{j=1}^{n} [f_j^b(x) - f_j^b(y)]| \leq \sum_{j=1}^{n} |f_j^b(x) - f_j^b(y)| < \varepsilon/M_i, \text{ on summing the}$$

inequalities (5.4) over $j = 1,\ldots,n$.

(iii) Consider a given translate aV. Part (i) gives $aV \tilde{G} aG_e$,

and since μ_i has $\tilde{G}_e$ as a right period, it is constant on the dense

subset $aV \tilde{G}$ of aV. Thus the continuous function μ_i^b is constant

on aV.

With Lemma 5.3 established, the rest of the proof of Theorem
4.5 directly generalizes the proof of (the existence part of) von
Neumann's theorem in [4, 41D]. The proof is based on the construction
of a "partition of unity", a finite set $\{g_1,\ldots,g_r\}$ of continuous non-
negative functions $g_k : \tilde{G}^b \to \mathbf{R}^+$ with a finite subset $\{a_1,\ldots,a_r\}$ of

$\tilde{G}$ such that g_k vanishes outside the translate $a_k V$ and $\sum_{k=1}^{r} g_k = 1$.

Taking $c_k = \int g_k(x)dx > 0$ with $\sum_{k=1}^{r} c_k = 1$, it will then be shown

that for y in $\tilde{G}^b$, the value $u^b(y)$ of the continuous function on

$\tilde{G}^b$ that restricts to the fixed solution u of $\Delta_i u = 0$ differs from

the convex combination $\sum_{k=1}^{r} c_k (T_i^{a_k} f)^b$ evaluated at y by less that
ε.

Towards the construction of the partition of unity, recall
that in the free group $\tilde{G}$, the intersection of all the subgroups K of
finite index is the identity element. Thus in $\tilde{G}^b$, the intersection of
all the closures $\bar{K}$ is the identity element. This means that $\tilde{G}^b$ is a
T_1-space, and hence Hausdorff [4, 28D]. Since $\tilde{G}^b$ is also compact, it
is normal [4, 3B], and thus Urysohn's Lemma applies [4, 3C]. Taking the
complement of V and the singleton $\{v\}$ of a point v in V as the

closed sets F_0, F_1, use Urysohn's Lemma to construct a continuous non-negative function $h : \tilde{G}^b \to \mathbf{C}$ vanishing outside V and with $h(v) = 1$. let U be the inverse image of the set of positive reals under h; then U is a non-empty open subset of V. The set $\{aU \mid a \in \tilde{G}\}$ of all translates of U covers the compact space $\tilde{G}^b$, and thus contains a finite subcover $\{a_1 U, \ldots, a_r U\}$. Then

$$\sum_{k=1}^{r} T^{a_k} h(x) > 0 \text{ for all } x \text{ in } \tilde{G}^b, \text{ so that } g_k = (T^{a_k} h)\left(\sum_{\ell=1}^{r} T^{a_\ell} h\right)^{-1}$$

is a well-defined continuous function on $\tilde{G}^b$ for $1 \leqslant k \leqslant r$. These functions g_k form the partition of unity.

The fixed solution u of $\Delta_i u = 0$ to be taken is $u = \mu_i * f / \mu_i(1)$. Recall the idempotence of μ_i under convolution demonstrated in Proposition 3.11. Moreover, as a consequence of the associativity of $*$ on $L^1(\tilde{G}^b)$ [4, 31B], the convolution on $AP(\tilde{G})$ is associative. Then $\mu_i * u = \mu_i * \mu_i * f / \mu_i(1) = \mu_i * f / \mu_i(1) = u$, so that $\Delta_i u = 0$.

Finally, consider an element y of $\tilde{G}^b$. Then

$$\left| u^b(y) - \sum_{k=1}^{r} c_k (T_i^{a_k} f)^b(y) \right|$$

$$= \mu_i(1)^{-1} \left| \mu_i^b * f^b(y) - \sum_{k=1}^{r} \int g_k(x) \mu_i^b(a_k) f^b(a_k^{-1} y) dx \right|$$

$$= \mu_i(1)^{-1} \left| \sum_{k=1}^{r} \int g_k(x) [\mu_i^b(x) f^b(x^{-1} y) - \mu_i^b(a_k) f^b(a_k^{-1} y)] dx \right|.$$

The k-th integrand is only non-zero where $g_k(x)$ is non-zero, i.e. for x within the translate $a_k V$. Since such x and a_k both lie within $a_k V$, Lemma 5.3(iii) shows that $\mu_i^b(x) = \mu_i^b(a_k)$. Further, since $(x^{-1} y)(a_k^{-1} y)^{-1} = x^{-1} a_k \in V^{-1} = V$, Lemma 5.3(ii) shows that $|f^b(x^{-1} y) - f^b(a_k^{-1} y)| < \varepsilon / M_i$. But $|\mu_i^b(x)/\mu_i(1)| \leqslant M_i$, so that

$$\left| u^b(y) - \sum_{k=1}^{r} c_k (T_1^{a_k} f)^b(y) \right|$$

$$= \left| \sum_{k=1}^{r} \int g_k(x) \frac{\mu_i^b(x)}{\mu_i(1)} \lceil f^b(x^{-1}y) - f^b(a_k^{-1}y) \rceil dx \right|$$

$$\leqslant \sum_{k=1}^{r} \int g_k(x) \left| \frac{\mu_i^b(x)}{\mu_i(1)} \right| \cdot \left| f^b(x^{-1}y) - f^b(a_k^{-1}y) \right| \, dx$$

$$< \sum_{k=1}^{r} \int g_k(x) M_i (\varepsilon/M_i) dx$$

$$= \varepsilon \int \left(\sum_{k=1}^{r} g_k(x) \right) dx = \varepsilon \int 1 dx = \varepsilon \, ,$$

as required to complete the proof of the theorem.

REFERENCES

[1] E. BANNAI and T. ITŌ, "Algebraic Combinatorics I: Association
Schemes", Benjamin-Cummings Mathematics Lecture Notes No. 58, Menlo
Park 1984.

[2] J. DIXMIER, Les C^*-Algèbres et leurs Répresentations, Gauthier-
Villars Cahiers Scientifiques No. 29, Paris 1964.

[3] K. W. JOHNSON and J. D. H. SMITH, Characters of finite quasigroups,
Eur. J. Comb. 5(1984), 43-50.

[4] L. H. LOOMIS, "An Introduction to Abstract Harmonic Analysis", Van
Nostrand, New York 1953.

[5] J. v. NEUMANN, Almost periodic functions in a group I, Trans. Amer.
Math. Soc. 36(1934), 445-492.

[6] J. D. H. SMITH, "Representation Theory of Infinite Groups and
Finite Quasigroups", Séminaire de Mathématiques Supérieures,
Université de Montréal, Montréal 1986.

GEOMETRIC METHODS IN GROUP THEORY

Stephen D. Smith[*]
Department of Mathematics
University of Illinois at Chicago
Chicago, IL 60680 USA

[*]Partially supported by the National Science Foundation.

Abstract. A survey of recent results which combine techniques from the areas of discrete geometry and finite group theory.

INTRODUCTION

Four years ago, some members of the present audience were also gathered in Montreal, for the 1982 meeting "Finite Groups: Coming of Age" organized by John McKay of Concordia University. At that time, I presented a survey lecture [Sm1] on the comparatively new and rapidly developing area of "groups and geometries". By now the area has come to be a more established branch of modern mathematics. It seems appropriate to make today's lecture in effect a sequel to that 1982 lecture - to describe further progress on the main problems then open and indicate important new directions which have opened up since that time.

It will be convenient to follow again the general outline of the earlier lecture, namely:

1) Motivation and applications.

2) Background on geometries and diagrams.

3) "Sporadic" geometries.

4) Properties and characterizations.

Of course, events have rendered parts of this organization a little outdated, but the parallel treatment should help emphasize developments since 1982.

1 MOTIVATION AND APPLICATIONS

At the earlier meeting, I presented "groups and geometries" as an area whose more widespread development had begun in the latter days (late 70's) of the classification of finite simple groups. In overview, that massive result shows that a non-abelian finite simple group must be one of:

> (a) an alternating group;
>
> (b) a group of Lie type, defined over a finite field;
>
> (c) one of 26 "sporadic" groups - not contained in the families (a) (b).

A principal motivation for the introduction of geometric techniques is the hope of obtaining a uniform understanding of the exceptional cases (c) alongside the generic case (b). I will begin with several recent results on subgroup structures, which seem to me to represent important progress in this direction.

For a Lie-type group over a field of characteristic p, the most important subgroups are the p-local subgroups (normalizers of p-subgroups); these are contained in the parabolic subgroups, whose geometry is described by the Tits building [Tits1]. It was observed in Ronan-Smith [RSm1] and Ronan-Stroth [RSt] that for many sporadic groups there is at least one prime p, such that the lattice of p-local subgroups exhibits a geometry with analogy to a building. More recently, Aschbacher [A1] has developed these earlier observations into an actual theory of subgroup structure. Briefly, Aschbacher writes down a short list of axioms for a set of "minimal p-parabolics"; these are easily verified for Lie-type groups, and verified individually for sporadic groups; he then demonstrates from the axioms that all subgroups above a Sylow-p-group are determined by the geometry defined by the generating parabolics.

I would also like to mention work of recent years on the maximal-subgroup problem for simple groups, which I believe can be regarded as geometric in overview. A common feature of modern approaches is to represent the desired simple group G as the automorphism group of some suitable object Δ; this might be a set when G is a permutation group, a vector space with a form when G is a matrix group, and so on. In any case, we might regard Δ and its relevant incidence structures as an abstract geometry. The aim of the approach is then to prove a "structure theorem" for G and Δ, asserting that a maximal subgroup M of G must satisfy:

> either (i)   M   is the stabilizer of some natural subconfiguration of Δ,
>
> or   (ii) $F^*(M)$ is simple, and acts irreducibly (in some sense) on Δ.

Such a result for symmetric groups was first established by O'Nan-Scott and Aschbacher [Sco,ASc]. Later, Aschbacher [A2] obtained the relevant

result for the classical matrix groups, in terms of the geometry of the
natural quadratic form. More recent work of Aschbacher [A3] aims to provide similar structure theorems for exceptional Lie-type groups (and at
least some sporadic groups) in terms of the geometry of a suitable set of
multilinear forms.

In 1982, I mentioned as a second likely by-product of the
geometric approach the possibility of producing simplifications in the
present classification proof for simple groups (the area of "revisionism"
led by Gorenstein). This expectation also appears to have been dramatic-
ally fulfilled. Stellmacher [Ste] is now in the process of applying the
essentially geometric "amalgam" method to shorten the quasithin-group
problem; the original classification of quasithin groups by Mason [M] in
about 600 journal pages represents one of the longest elements of simple-
group classification, and the last to be published. A second instance:
Stroth [Str1] is using related techniques in a new approach to the "uniqueness" problem – the lengthy original work by Aschbacher [A4] showed
essentially that a certain simple group could not have strongly p-
embedded 2-locals, and thus established the final contradiction completing simple-group classification.

Furthermore, it seems reasonable to expect that the geometric
viewpoint will be used more fully in revising those parts of the classification where it is necessary to recognize (i.e., characterize uniquely)
some group by means of given p-local subgroups. Details should emerge
as the work of Gorenstein, Lyons, R. Solomon, _et al_. progresses.

2 BACKGROUND ON GEOMETRIES FOR GROUPS

This section is intended to provide the non-specialist with a
quick introduction to the notions of abstract geometry and group actions.
For fuller details, the reader may wish to consult the earlier article,
or any of a number of other expository articles which have appeared in
recent years (e.g. [Tits3]).

Geometries

We can begin with the example of projective space, which most
mathematicians have at least encountered. We let V denote an n-dimensional vector space over a field $\mathbb{F}_q$, with G denoting its full group
of linear automorphisms $GL(V)$. Then the projective space $\mathbb{P}V$ has as
its objects (or varieties) the linear subspaces of V:

projective points   = linear 1-subspaces;

projective lines   = linear 2-subspaces;

$\vdots$ $\vdots$

projective hyperplanes = linear $(n-1)$-subspaces.

These objects may furthermore be related by inclusion; a chain of subspaces so related is called a _flag_, and the size of a maximal such chain
($n-1$, in this example) is the _rank_ of the geometry. Evidently G permutes the basic objects, and preserves the inclusion relation - and so
also permutes the flags. Indeed the fundamental theorem in linear algebra, that any two bases are related by an invertible matrix, implies the
statement that G is _flag-transitive_ (that is, has just one orbit on
the maximal flags).

The reader can easily visualize similar geometries. For
example, if V is equipped with a quadratic form, one focuses attention
only on the spaces isotropic in the form, and for G uses just the corresponding orthogonal group. More generally we need not restrict attention to subspaces of a vector space V; an abstract _geometry_ will simply
be some set, usually divided into objects of various "types" (such as
points, lines, etc. above), related by some notion of _incidence_. The
notions of flag, rank, and type-preserving automorphism carry over. The
relevance of these ideas to group theory, and particularly to simple
groups, is that a geometry with a high degree of symmetry is likely to
have a flag-transitive automorphism group - and this high transitivity
is a typical feature of a simple automorphism group. In particular, the
suggestion arises that one might approach simple groups in a _uniform_
manner by the study of actions on geometries.

As an example in which the geometry may not be initially
evident, consider the Mathieu group M_{24}. Ordinarily this group is
viewed either as a special 5-transitive permutation group, or the automorphism group of the extended binary Golay code, or the automorphism
group of the Steiner system $S(5,8,24)$. The latter view leads to a
natural geometry; we use the terminology of Conway [Co], to which the
reader is referred for fuller details. The Steiner system is a special
collection of 759 subsets of size 8, called _octads_, chosen from a fixed
24-set. A triple of octads partitioning the 24-set is called a _trio_. A
partition of 24 into 6 sets of size 4 is called a _sextet_ if each pair of
4-sets in the partition gives one of the special octads. Now the collec-

tion of all octads, trios and sextets (with the obvious inclusion relation) forms a very natural rank-3 geometry, on which M_{24} in fact acts
flag-transitively. Indeed, this is a _2-local_ geometry in the sense of
Ronan-Smith [RSml], since the subgroups stabilizing the geometric objects
are in fact the maximal 2-constrained 2-local subgroups of M_{24}. (It
happens that this geometry can be embedded as certain subspaces of $\mathbb{P}V$
for V of dimension 11 over $\mathbb{F}_2$).

Buildings and diagrams

The most impressive use of the geometric approach is the work
of Tits [Tits1], which showed that the groups of Lie-type act on geometries which can be uniformly axiomatized - the theory of buildings.

As background, we recall briefly the fundamental construction
of Chevalley (and Steinberg's "twisted" variation), which showed that
the groups of Lie-type really are the "generic" family of simple groups.
We fix one of the complex simple Lie algebras L, and its Weyl group W.
Then given a field $\mathbb{F}_q$, the construction produces (as a linear group
over $\mathbb{F}_q$) a simple group said to be of "type L". One thus obtains the
classical matrix groups, as well as the exceptional types G_2, F_4, E_6,
E_7, E_8. The twisted types, such as unitary groups, arise as fixed points
in these groups of certain outer automorphisms.

Tits' work arose in the attempt to find an analogue for the
exceptional types of the obvious natural projective geometries for the
classical matrix groups. We will describe the work only in brief overview. Tits focused on the natural geometry for the Weyl group W - the
Coxeter complex, which is a triangulation of a suitable sphere. He then
was able to formulate the axioms for a _building_, which in particular can
be regarded as a collection of such spheres ("apartments"), with suitable
boundary identifications. Tits' main result is that a building of rank
≥ 3 must arise from a Lie-type group - and thus buildings are precisely
the natural geometries for this generic class of simple groups.

The building viewpoint leads naturally to the association of
a _diagram_ to a geometry. With a Weyl group W there is associated the
Coxeter-Dynkin diagram describing the generation of W by certain in-
volutions. But in fact the diagram also contains some information on
the particular axioms satisfied by the geometry. Recall for instance
the earlier example of projective space. The group $GL(V)$ (or more
precisely, its simple section $PSL(V)$) arises in Chevalley's construc-

tion from the Lie algebra $L \simeq s\ell_n$ of type A_{n-1}, with Weyl group $W \simeq S_n$. The Dynkin diagram is linear:

$$
\begin{array}{cccccccc}
1 & 2 & 3 & 4 & & n-3 & n-2 & n-2 \\
\circ\!-\!\!-\!\circ\!-\!\!-\!\circ\!-\!\!-\!\circ & & \cdots & & \circ\!-\!\!-\!-\!\circ\!-\!\!-\!-\!\circ
\end{array}
$$

Here the numbering of the nodes corresponds to the linear dimensions of the various subspaces of V, and structural information can be deduced (inductively), for example: Suppose we fix some 3-dimensional subspace W of V; the <u>residue</u> of W is the subgeometry of all objects incident to W. Correspondingly we remove the 3-dimensional node from the diagram to obtain

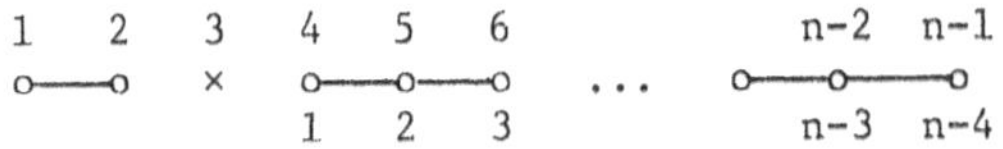

$$
\begin{array}{cccccccc}
1 & 2 & 3 & 4 & 5 & 6 & & n-2 & n-1 \\
\circ\!-\!\!-\!\circ & & \times & \circ\!-\!\!-\!\circ\!-\!\!-\!\circ & & \cdots & & \circ\!-\!\!-\!\circ\!-\!\!-\!\circ \\
& & & 1 & 2 & 3 & & n-3 & n-4
\end{array}
$$

The diagram remaining expresses the fact that the residue of W is the geometric product of a projective plane ($\overset{1}{\circ}\!-\!\!-\!\overset{2}{\circ}$) for the subspaces inside W, with another projective space ($\underset{1}{\circ}\!-\!\!-\!\underset{2}{\circ}\!-\!\!-\!\underset{3}{\circ} \ \cdots \ \underset{n-3}{\circ}\!-\!\!-\!\underset{n-4}{\circ}$) for the spaces above W (that is, the projective space of the quotient V/W).

If we release the building axioms, and hence the tight connection with the Weyl group W, the above residue conventions will define inductively a weaker notion of a geometry "belonging to a diagram", provided we fix xonventions for rank-2 diagrams; typically one uses the standard generalized polygons:

$$
\begin{array}{cccc}
\circ \quad \circ & \circ\!-\!\!-\!\circ & \circ\!=\!\!=\!\circ & \circ\!=\!\!=\!\circ
\end{array}
$$

generalized	generalized	generalized	generalized
digon	triangle	quadrangle	hexagon
(i.e., product)	(i.e., projective plane)		

In fact, this viewpoint was a component of Tits' original approach to buildings, and is adopted in his more recent and very influential paper [Tits2].

As a case of the association of a diagram to a non-building geometry, recall the earlier example of M_{24}. Here the diagram in the conventions of [RSm1] is:

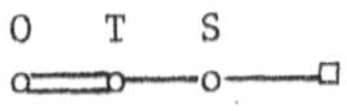

$$
\begin{array}{ccc}
O & T & S \\
\square\!=\!\!=\!\square\!-\!\!-\!\circ\!-\!\!-\!\square
\end{array}
$$

Thus, the residue of an octad (o——o——□) comes from $\mathbb{P}V$ for a space
                                1   2
V of dimension 4 (over $\mathbb{F}_2$) – but the square node tells us to ignore
the linear spaces of dimension 3, and use only those of dimensions 1 and
2. Similarly, the residue of a sextet (o≡≡o) is the generalized quad-
                                        1   2
rangle for a symplectic 4-space over $\mathbb{F}_2$.

    With these notions in hand, we can begin to survey develop-
ments in the geometric approach to simple groups.

3 GROUP GEOMETRIES – EXAMPLES AND CLASSIFICATION RESULTS

    Next we will briefly recapitulate (under several headings)
developments up to the time of [Sm1], and then indicate progress during
the intervening years. But first an overall historical note:

    Following the work of Chevalley-Steinberg and Tits mentioned
above, it was Buekenhout [Bu] who first began to extend the geometric
notions surrounding buildings to more general diagram geometries, especially for sporadic simple groups. This work did not immediately attract
the attention of group theorists in general, but its influence began to
show more and more in the following years. This was indicated in Tits'
local-approach paper [Tits2] developing the notion of a "geometry of type
M", which in turn has profoundly influenced subsequent developments. The
status at the time of the 1978 Durham conference is described in Tits'
survey article [Tits3]. In my 1982 article, I had indicated the discovery of a considerable variety of examples; it now remains to discuss
recent progress towards understanding and classification of those discoveries.

Amalgams and the amalgam method

    The body of work begun in Goldschmidt's "amalgams" paper
[Go] was mentioned only in passing in the earlier survey, because at that
time it seemed to represent a rather different direction from the geometric approach described in [Sm1]. But one development of the intervening
years was the refinement of the "amalgam method" into a tool of very
general applicability; and in particular it has been used very success-
fully in more purely geometric problems. Consequently, some treatment
is called for in the present article.

    The amalgam problem investigates in general a situation that
holds in particular in rank-2 Lie-type groups in some fixed characteris-

tic p: consider a group G generated by two finite subgroups P_1 and P_2 such that:

(i)   For i = 1,2 the group $\overline{P}_i := P_i/O_p(P_i)$ is a rank-1 Lie-type group in characteristic p;

(ii)   The intersection $B := P_1 \cap P_2$ is the normalizer of a common Sylow p-group of P_1 and P_2 (but contains no normal subgroup of G).

The group G itself can in general be any quotient of the (infinite) amalgamated product $P_1 *_B P_2$, so one cannot hope for a detailed description of G. What is astonishing is that it is possible, beginning only with the structure of the quotients $\overline{P}_i$, to list all possibilities for the structure of the generating subgroups P_i!

The geometric content arises by considering the cosets of P_1 and P_2 – these define a graph on which G acts edge-transitively. The structure of $\overline{P}_i$ essentially describes just the local action of G – that is, how P_i acts on the "neighboring" cosets of the other subgroup (those which intersect it). The interest of the situation becomes apparent when we observe that such amalgams arise not only for rank-2 Lie-type groups, but also for such sporadic groups as M_{12}, J_2, J_3, and F_3.

The first case, with $\overline{P}_i \simeq GL_2(\mathbb{F}_2)$, was handled by Goldschmidt [Go]. The results were then extended by the work of various others (including Stroth and Fan), culminating in the general classification of Delgado, Goldschmidt and Stellmacher [DGS], to which the reader is referred for fuller details and references.

This work was nearly complete by 1982, but there is much to add since then. Beyond the utility and significance of the result itself, events have shown that the amalgam method is useful in other situations. Much of this is due to Stellmacher, whose analysis of Goldschmidt's work made fully explicit the use of weak closure methods, so that it was no longer necessary to make an elaborate study of "tracks" in the graph. Further advances resulted when other authors such as Stroth [Str2] and Timmesfeld [Tim1] were able to use this weak closure-based approach in more purely geometric classification problems. And in "revisionism", the techniques have been applied (as mentioned earlier) by Stroth and Stellmacher.

General Diagram Geometries

In associating diagrams to geometries as described above,
one direction to follow is to create suitable new rank-2 residues, in
addition to the conventional generalized n-gons. The motivation for
this step is not esthetic but entirely ad hoc: in studying the sporadic
groups, it quickly becomes clear that other rank-2 geometries do arise in
practice. The "circle" geometries of Buekenhout [Bu], and the "square-
node" truncations appearing in Ronan-Smith [RSm1] appear to have been
particularly influential.

In view of the highly specific origin of these notions, it
does not seem realistic to expect classification theorems for geometries
with thse more general diagrams - and indeed I know of no general results
in such a direction since 1982. Nevertheless, various more particular
results have emerged, and I will mention several of these. For diagrams
with square nodes, there are certain criteria which guarantee that the
geometry in fact belongs to the corresponding filled-in diagram, with
all square nodes replaced by ordinary nodes (often, a building). Results
of this type were established by Ronan [R1], and similar results by
Brouwer-Cohen [BC1]. For particular diagrams, it is also sometimes possible to show uniqueness - namely that only one group (or class of
groups) can belong to the diagram. Such proofs for the M_{24} 2-local
diagram were mentioned already in [Sm1]. In that article, a brief reference was made to the possibility of a similar uniqueness proof for the
Conway group $\cdot 1$ (with diagram ⊏══○────○────○ given in [RSm1]). Recently

Segev [Se] and Hewitt [He] have by different approaches produced such
uniqueness results. Furthermore, Hewitt is trying to carry the analysis
forward to a uniqueness proof for the next obvious candidate, the Monster
F_1 (with diagram ⊏══○───○───○───○ in [RSm1]).

In another direction, the overgroup theory of Aschbacher [A1]
mentioned earlier has clarified and made precise some of the applications
of more general geometric diagrams.

Tits Geometries (affine)

One of the many influential ideas in Tits' local approach
paper is that of "geometries of type M" (where M denotes a generalized

Cartan matrix). This class is defined by diagrams as described earlier, with the restriction that all rank-2 residues corresponding to generalized n-gons for some n. (The cases $n = 2,3,4,6$ mentioned earlier are the most common, as is more or less explained by the Feit-Higman theorem [FH]). This class is extremely interesting - it includes of course all the buildings and a considerable collection of other intriguing examples. Terminology for this class is not standard in the literature: Tits' term "type M" has not caught on; Kantor used the term GAB for "geometries that are almost buildings" [K1]; Timmesfeld and others have used "Tits geometries". Without wishing to detract from the honor due to Tits, I have tried (without much success) to popularize the term "generalized polytopes", in view of the generalized polygons from which they are inductively built up.

The status of the study of these geometries in 1982: Tits had established [Tits2] two important results: first, that the universal cover of such a geometry without C_3 or H_3 residues must in fact be a building; and a classification of those buildings of rank ≥ 4 whose diagrams correspond to one of the affine (or "Euclidean") Weyl groups. In addition, a bewildering variety of new geometries in this class had been found, notably in [RSm1], [K1]; the work of Timmesfeld [Tim2] on single-bond geometries led to the discovery of other examples [ASm, R2, KMW]. The years since 1982 have seen considerable progress in the understanding of this class.

First we consider the diagrams of affine type. One of the most satisfying results is that of Kantor [K2]. The geometry with affine diagram $\tilde{D}_4$(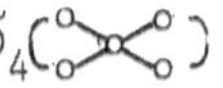) described in [ASm] is studied by passing to the universal cover as suggested by Tits - indeed the affine building is identified as that of the group of type D_4 over the 2-adics $\mathbb{Q}_2$. Consideration of automorphisms and fixed sublattices leads to a similar "explanation" of most of the known examples of types $\tilde{B}_2$(), $\tilde{C}_2$(), $\tilde{G}_2$(), and $\tilde{A}_3$(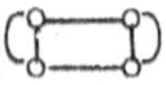). In fact, Kantor's work not only determines the universal cover of these examples - it also shows, via lattice quotients, that each of those examples is just the smallest member of an infinite sequence of finite examples.

As Kantor's work was being formulated, further examples of various diagram geometries were being discovered, notably by Köhler, Meixner and Wester in Giessen. (A general method of construction is described by Meixner in [Me]). They were then quick to interpret their examples in the light of Kantor's universal approach [MW]. For further examples, and perhaps the most complete list of references, the reader should consult Wester's thesis [Ws]. Other contributions to this area were made by Tits [Tits4].

One particular diagram not covered in Kantor's approach is that of type $\tilde{A}_2$ (⟨triangle diagram⟩). These geometries also first arose in Timmesfeld's single-bond work [Tim2]; further examples were discovered, and their universal covers investigated, by Ronan [R2], van Maldeghem [V], and Kohler-Meixner-Wester [KMW].

After all these decisive results, it is necessary to report that apparently no progress has been made on the rank-3 affine buildings associated to sporadic groups: that of type $\tilde{C}_2$ (○═══○═══○) for Suz, and that of type $\tilde{G}_2$ (○────○═══○) for Ly. The nature of their universal covers seems to remain a mystery.

Tits Geometries (general case)

It remains now to describe some classification results for the more general category of generalized polytopes - where the diagram need not be assumed to be of affine type. We restrict attention for a moment to the case of spherical diagrams. In the local approach paper [Tits2], Tits had shown that any geometry belonging to type A_n ($n \geq 3$) must be a building, and gave sufficient extra conditions for the other diagram types to guarantee the building axioms. More recently, Timmesfeld [Tim2] (followed by Brouwer-Cohen [BC2]) observed that the above extra conditions are unnecessary for types D_n and E_6 - that such geometries must in fact be buildings. Also Aschbacher [A5], assuming spherical diagram and natural group theoretic hypotheses, establishes that the only possibilities are the usual buildings, and the sporadic C_3 geometry (for the alternating group A_7 first discovered by Neumaier [N]). This result has been quoted in much subsequent work on group geometries.

Other classification efforts have added natural group-theoretic hypotheses to the geometric assumptions for generalized polyhedra. This area had been started, at the time of [Sm1], by work of Timmesfeld [Tim2], which considered diagrams with single bonds, where the projective-plane residues are assumed to be Desarguesian; he showed the only cases to arise are the usual buildings, plus examples of types $\tilde{A}_2$,

$\tilde{A}_3$, and $\tilde{D}_4$ (indeed, the discovery of these examples was a by-product of Timmesfeld's work). Not long thereafter, Timmesfeld was able to extend his analysis of the case A_2 (△) to a triangle diagram with bonds of any strength – the result [Tim4] essentially showing that any rank-3 geometry with classical residue groups must have a linear diagram. (Very recently, Timmesfeld has announced [Tim3], a generalization which depends only on rank-1 residues). This result has been crucial in simplifying the analysis of further diagram geometries. In a series of papers, Stroth [Str2] has essentially extended Timmesfeld's result to full generality by allowing double bonds; the list of possibilities includes the buildings plus a larger collection of examples, some of which can be described only locally (as in the $\tilde{A}_2$, $\tilde{A}_3$, $\tilde{D}_4$ cases of Timmesfeld). Of course, an added complication in Stroth's work is the possibility that Neumaier's C_3 geometry can occur as a rank-3 residue.

4 SOME SPECIAL GEOMETRIC PROPERTIES

Once geometries such as those of the previous section have been discussed, the desirability of corresponding classification results is at once evident. Of course, another part of the understanding of the discoveries consists in a study of their distinctive properties; here I will continue the discussion in [Sm1] of various combinatorial-topological features of these new geometries.

With regard to geometries belonging to affine diagrams, the previous section indicated results determining the structure of universal covers. Many of the geometries for other diagrams are now known to be simply-connected (and even to have the homotopy Cohen–Macaulay property as mentioned in [Sm1]). But I believe it is fair to say that these have been <u>ad</u> <u>hoc</u> calculations, and that there has been no uniform treatment of simple connectivity.

Shellability

The earlier survey also mentioned the combinatorial notion of shell-ability as a possible direction for analysis. Developments in the intervening years have seemed to indicate that this property is relevant for buildings (as observed by Björner [Bj1]) but does not cast light on the more general geometrics. For example, the notion of metrical shell-ability developed by Scharlau (see [Sch]) turns out to lead to the usual formulation of buildings. The particular case of the sporadic C_3 geometry for A_7 (mentioned earlier) may yet be interesting. At the time of [Sm1], the geometry had been shown by Ronan to be simply-connected, and so to have the homotopy Cohen-Macaulay property; but it appeared not to have the stronger property of shellability. Later I was able to verify (unpublished) that it does not have the still stronger property of CL-shellability (see [Bj1]). Walker [Wa] has directly constructed a complex which is shellable but not CL-shellable; I wonder if this C_3 geometry might in fact represent an example occurring "in nature". One way of demonstrating shellability would be to find a set of just 56 out of the 315 maximal faces, whose removal would leave the remaining complex topologically collapsible. The closet I have come to date is collapsibility on removal of a certain set of 61 faces, which seems tantalizingly close. But the shellability problem remains open.

Recently I received from Björner [Bj2] a general survey of Cohen-Macaulay results arising in group-complexes, to which the reader is also referred.

Homology and Projective Modules

The earlier article also mentioned briefly the representation space for G afforded by the highest dimensional homology of a geometry Δ. There has been a dramatic improvement of our understanding of phenomena here, and the topic seems especially relevant for mention at this combinatorial conference. (I gave a somewhat fuller exposition [Sm2] at the Bruck conference in Madison in July 1985, and refer the reader for fuller details to that article, or the original references below).

Once again the generic case providing the motivation is that of Lie-type groups and buildings:

> <u>Theorem</u> (L. Solomon-Tits [So]). Let G be a finite Lie-type group of rank n and characteristic p, and Δ its building. Then the top homology group $H_{n-1}(\Delta)$ with coefficients in $\mathbb{F}_p$ affords the Steinberg $\mathbb{F}_p G$-module.

This important module is distinguished among the irreducible $\mathbb{F}_p G$-modules in that is also <u>projective</u>. (For the reader unfamiliar with the theory of projective modules, we will in fact need here only the property that such a module has dimension divisible by the p-part $|G|_p$ of the group order; see a standard reference such as [Do]).

It is then natural to ask if a similar phenomenon occurs for sporadic groups and their geometries. In a few examples, top homology does provide a projective (though not usually irreducible) module: the calculations for the C_3 geometry of A_7 and for the 2-local geometry of M_{24} were carried out by Ronan-Smith with Peter Webb (see [RSm2]). However, in typical sporadic examples, the module is not in fact project-ive; the necessary computations for a number of geometries appear in Ronan [R3]. For later reference, we will indicate a small case:

> Example: Let G be the Mathieu group M_{12}. Define a rank-2 geometry Δ as follows: <u>Points</u> are given by the various 4-subsets of the set of 12 letters on which M_{12} acts. <u>Lines</u> are given by that M_{12} conjugacy class of $4|4|4$ partitions whose stabilizers con-tain a Sylow 2-group. One computes that the top homology $H_1(\Delta)$ has dimension $496 = 2^4 \cdot 31$ – and so cannot be a projective $\mathbb{F}_2(G)$-module, since $|G|_2 = 2^6$.

Nonetheless, one observes that the power 2^4 is still a high one; and one is tempted to feel that the failure of projectivity is somehow not too severe. Similar observations were made in numerous other examples, and the explanation of this phenomenon represented an outstanding open problem.

The explanation which has emerged exhibits an intriguing in-teraction with the combinatorics of finite groups studied by topologists. Now we let G be any finite group, and p a prime dividing its order. The complex Δ_p is defined by taking as vertices the non-trivial p-subgroups of G, with higher-dimensional simplexes given by chains under inclusion. Then Δ_p determines via its chain groups $C_i(\Delta_p)$ and homo-logy groups $H_i(\Delta_p)$ standard invariants:

$$\tilde{L}(\Delta_p) := \sum_{i=-1}^{\dim \Delta_p} (-1)^i C_i(\Delta_p), \text{ the reduced \underline{Lefschetz} (virtual)} \text{ \underline{module}};$$

$$\tilde{\chi}(\Delta_p) := \sum_{i=-1}^{\dim \Delta_p} (-1)^i \dim(C_i(\Delta_p)), \text{ the reduced \underline{Euler charact-} \underline{eristic}.}$$

By the Hopf trace formula, the above L has the same character as the usual Lefschetz module defined via homology. Since the complex Δ_p is ordinarily connected, we have inserted a 1-dimensional term $H_{-1}(\Delta_p)$ to cancel the corresponding 1-dimensional term $H_0(\Delta_p)$. The result which initiated this area of study was:

> <u>Theorem</u> (Brown [Br]). $\tilde{\chi}(\Delta_p) \equiv 0 (\mathrm{mod}\,|G|_p)$.

A module-theoretic exppanation of this fact was given by the celebrated work of Quillen:

> <u>Theorem</u> (Quillen [Q]). $\tilde{L}(\Delta_p)$ is projective (virtual)
> $\mathbb{F}_p G$-module.

This work also established a connection with the Solomon-Tits theorem above, via:

> <u>Theorem</u> (Quillen [Q]). If G is a Lie-type group in char-
> acteristic p, and Δ is homotopy-equivalent to Δ_p. In
> particular, $\tilde{L}(\Delta_p)$ affords the Steinberg module.

It is relevant to remark that the complex Δ_p of all p-subgroups is much larger and more complicated than the building complex Δ, so the equivalence represents a considerable simplification. Evidently Quillen's result also provides projective modules from the complexes Δ_p for sporadic groups; again it is typically the case that Δ_p is much more complicated than the natural geometries Δ defined by p-local structure - and it is precisely the above homotopy-equivalence which often fails in sporadic cases.

Peter Webb subsequently studied Quillen's work with the intention of extending the result to more general group actions on complexes. He established:

> <u>Theorem</u> (Webb [We]). Suppose G acts on a complex Δ, and
> assume:
> For all non-trivial p-subgroups P of G, the fixed sub-
> complex Δ^P is contractible.
> Then $\tilde{L}(\Delta)$ is projective.

Later Thévenaz refined the result still further:

<u>Theorem</u> (Thévenaz [Th]). Suppose G acts on a complex Δ,
and assume:

X is a collection of p-subgroups, closed under subgroups and
conjugacy; for all $H \leq G$ with $H/O_p(H)$ cyclic and
$O_p(H) \notin X$, $\widetilde{\chi}(\Delta^H) = 0$.

Then $\widetilde{L}(\Delta)$ is projective relative to X.

For the reader unfamiliar with the notion of relative projectivity, we
simply require the consequence that if X is a member of X of largest
possible order, then $\widetilde{\chi}(\Delta) \equiv 0$ (modulo $|G|_p/|X|$).

Now Thévenaz' result enables us to explain near-projectivity
for many sporadic geometries. Let us return for example to the case of
M_{12}. Here there are two conjugacy classes of elements of order 2, with
representatives z and t; one check that Δ^z is contractible (a tree),
while Δ^t is not even connected. There are in fact groups T of order
4 all of whose non-trivial elements are conjugate to t; while any 2-
group P of order at least 8 contains a conjugate of z and in fact
satisfies Δ^P contractible. Consequently in applying Thevenaz' theorem
we may take X to be the conjugates of $\langle t \rangle$ and T, and conclude
$\widetilde{\chi}(\Delta) \equiv 0$ (modulo $2^{6-2} = 2^4$) - as needed. In general, we see that
Thevenaz' result provides method of computing a lower bound for the p-
part of $\widetilde{\chi}(\Delta)$; and in fact this bound is exact in the cases computed so
far.

A Note on Presheaf Homology

A generalization of ordinary homology also well-known to
topologists is that of presheaf homology - where one can assign to the
various simplexes not just a copy of the field, but arbitrary vector
spaces. Homology of the resulting complexes extends the applicability of
geometric techniques far more deeply into group representation theory.
The latter topic is perhaps less relevant to this conference - and would
in any case involve another survey lecture. (As some members of the
audience will know, I surveyed various results at the Bruck meeting in
Madison last year [Sm2], and further topics more recently, at the AMS
Summer Institute on Representation Theory in Arcata, to appear in those
proceedings [Sm3].

Acknowledgement

This article was prepared while the author was visiting the
University of Giessen. It is a special pleasure to thank Ingrid Ohm,
who volunteered to do the typing.

References

[A1]   Aschbacher, M. (1986). Overgroups of Sylow p-groups in sporadic
         groups. Memoirs A.M.S., no. 343; Providence, R.I., Amer.
         Math. Soc.
[A2]   Aschbacher, M. (1984). On the maximal subgroups of the finite
         classical groups. Inv. Math., $\underline{76}$, 469-514.
[A3]   Aschbacher, M. (1985). Some multilinear forms with large iso-
         metry group. Preprint, Caltech. Also, Chevalley groups of
         type G_2 as the groups of a trilinear form. Preprint,
         Caltech, 1985.
[A4]   Aschbacher, M. (1983). The uniqueness case for finite groups.
         Ann. Math., $\underline{117}$, 383-454, 455-551.
[A5]   Aschbacher, M. (1984). Finite geometries of type C_3 with flag-
         transitive groups. Geom. Dedic., $\underline{16}$, 195-200.
[ASc]   Aschbacher, M. & Scott, L. (1985). Maximal subgroups of finite
         groups. J. Algebra, $\underline{92}$, 44-80.
[ASm]   Aschbacher, M. & Smith, S. (1983). Tits geometries over GF(2)
         defined by Chavalley groups over GF(3). Comm. in Alg., $\underline{11}$,
         1675-1684.
[Bj1]   Björner, A. (1984). Some combinatorial and algebraic properties
         of Coxeter complexes and Tits buildings. Adv. in Math., $\underline{52}$,
         173-212.
[Bj2]   Björner, A. (1987). Some Cohen-Macaulay complexes arising in
         group theory. Advanced Studies in Pure Math., $\underline{11}$, 13-
         19.
[BC1]   Brouwer, A. & Cohen, A. (1986). Local recognition of Tits geo-
         metries of classical type. Geom. Dedic., $\underline{20}$, 181-200.
[BC2]   Brouwer, A. & Cohen, A. (1983). Some remarks on Tits geometries.
         Inv. Math., $\underline{45}$, 393-402.
[Br]   Brown, K.S. (1975). Euler characteristics of groups: the p-
         fractional part, Inv. Math., $\underline{29}$, 1-5.
[Bu]   Buekenhout, F. (1979). Diagrams for geometries and groups.
         J. Com. Th. A., $\underline{27}$, 121-151.
[Co]   Conway, J.H. (1971). Three lectures on exceptional groups.
         Pp 215-247 in "Finite Simple Groups", eds. Powell & Higman.
         NY: Academic Press, 1971.
[DGS]   Delgado, A., Goldschmidt, D., & Stellmacher, B. (1985). Groups
         and Graphs. Basel: Birkhaüser.
[Do]   Dornhoff, L. (1972). Group Representation Theory. NY: Marcel
         Dekker.
[FH]   Feit, W. & Higman, G. (1964). The nonexistence of certain gen-
         eralized polygons. J. Alg., $\underline{1}$, 114-131.
[Go]   Goldschmidt, D. (1980). Automorphisms of trivalent graphs. Ann.
         Math., $\underline{111}$, 377-406.
[He]   Hewitt, P. (a characterization of ·1 - work in progress).
[K1]   Kantor, W. (1981). Some geometries that are almost buildings.
         Europ. J. Comb., $\underline{2}$, 239-247.

[K2]   Kantor, W. (1985). Some exceptional 2-adic buildings. J. Alg.,
        92, 208-223. Some locally finite flag-transitive buildings.
        Europ. J. Comb., to appear.

[KMW]   Kohler, P., Meixner, Th. & Wester, M. (1984). Triangle groups.
        Comm. in Alg., 12, 1595-1626. The affine building of type

        $\tilde{A}_2$ over a local field of characteristic two. Arch. Math.,

        42. (1984), 400-407. Exceptional systems for the alternat-
        ing group A_7. Mitt. Math. Seminar Giessen, 165, (1984),
        19-34.

[M]   Mason, G. Quasithin groups. To appear in Memoirs A.M.S.

[Me]   Meixner, Th. (1984). Chamber systems with extended diagram.
        Mitt. Math. Seminar Geissen, 165, 93-104.

[MW]   Meixner, Th. & Wester, M. (1986). Some locally finite buildings
        derived from Kantor's 2-adic Groups. Comm. In Alg., 14,
        389-410.

[N]   Neumaier, A. (1984). Some sporadic geometries related to
        PG(3,2). Arch. Math., 42, 89-96.

[Q]   Quillen, D. (1978). Homotopy properties of the poset of non-
        trivial p-subgroups. Adv. Math., 28, 101-128.

[R1]   Ronan, M. (1986). Extending locally truncated buildings and
        chamber systems. To appear in Proc. London Math. Soc., 83.

[R2]   Ronan, M. (1984). Triangle geometries. J. Comb. Th. A., 37,
        294-319.

[R3]   Ronan, M. (1981). Coverings of certain finite geometries. Pp.
        316-331 in Finite Geometries and Designs (eds. Cameron,
        Hirschfeld, Highes). London Math. Soc. Lecture Notes No. 49.
        Cambridge, Cambridge U. Press.

[RSm1] Ronan, M. & Smith, S. (1979). 2-Local geometries for some spor-
        adic groups. Pp. 283-289 in Proc. Symp. Pure Math. No. 37
        (Santa Cruz 1979 Proceedings, eds. Copperstein & Mason).
        Providence, R.I., Amer. Math. Soc., 1980.

[RSm2] Ronan, M. & Smith, S. (1985). Computation of 2-modular sheaves
        and homology for $L_4(2)$, A_7, $3S_6$ and M_{24}. Preprint, Univ.
        of Ill.-Chicago.

[RSt]   Ronan, M. & Stroth, G. (1984). Minimal parabolic geometries for
        sporadic groups. Europ. J. Comb., 5, 59-91.

[Sch]   Scharlu, R. (1985). Metrical shellings of simplical complexes.
        Europ. J. Comb., 6, 265-269. A characterization of Tits
        buildings by metrical properties. J. London Math. Soc., 32,
        317-327.

[Sco]   Scott, L. (1979). Representations in characteristic p. Pp.
        319-332 in Proc. Symp. Pure Math. No. 37 (Santa Cruz 1979
        Proceedings, eds. Cooperstein & Mason). Providence, R.I.:
        Amer. Math. Soc., 1980.

[Se]   Segev, Y. (a characterization of ·1 - work in progress).

[Sm1]   Smith, S. (1982). Residual geometries for sporadic and classi-
        cal groups - a survey. Pp. 315-334 in Contemporary Mathema-
        tics, vol. 45. (Montreal 1982 Proceedings, ed. John McKay)
        Providence, R.I., Amer. Math. Soc., 1985.

[Sm2]   Smith, S. (1985). Homology representations from group geome-
        tries. Algebras, Groups & Geometries, 2, 514-540.

[Sm3]   Smith, S. (1987). Geometric Techniques in Representation Theory.
        To appear in Proc. Symp. Pure Math. (ed. P. Fong). Provi-
        dence, R.I.: Amer. Math. Soc., 1987.

[So]   Solomon, L. (1969). The Steinberg character of a finite group
         with a BN-pair. Pp. 213-221 in Theory of Finite Groups (eds.
         Brauer & Sah). NY: Benjamin, 1969.
[Ste]   Stellmacher, B. Work in progress on quasithin groups.
[Str1]   Stroth, G. Work in progress on the uniqueness problem.
[Str2]   Stroth, G. (1985). Chamber systems, geometries, and parabolic
         systems whose diagram contains only bonds of strength 1 and
         2. Preprint #205, F.B. Math., F.U. Berlin, 1985. A local
         classification of finite classical Tits geometries of char-
         acteristic not 3. Preprint #208, F.B. Math., F.U. Berlin,
         1986.
[Th]   Thévenaz, J. (1985). Permutation representations arising from
         simplicial complexes. Preprint, École Normale Supérieure,
         Paris.
[Tim1]   Timmesfeld, F.G. (1984). On the non-existence of certain flag-
         geometries. Mitt. Math. Sem. Giessen, 163, 19-44.
[Tim2]   Timmesfeld, F.G. (1983). Tits geometries and parabolic systems
         in finitely generated groups. Math. Z., 184, 377-396 and
         449-487.
[Tim3]   Timmesfeld, F.G. (1986). On amalgamation of rank-1 parabolic
         groups. Preprint, U. Giessen.
[Tim4]   Timmesfeld, F.G. (1984). Locally finite classical Tits chamber
         systems of rank 3 and characteristic 2. Preprint, U.
         Giessen.
[Tits1] Tits, J. (1974). Buildings of spherical type and finite BN-
         pairs. Lecture Notes in Math., No. 386. Berlin-NY,
         Springer-Verlag.
[Tits2] Tits, J. (1981). A local approach to buildings. Pp. 519-547 in
         The Geometric Vein (Coxeter Festschrift, eds. Davis, Grun-
         baum, & Scherk). NY: Springer-Verlag.
[Tits3] Tits, J. (1980). Buildings and Buekenhout geometries. Pp. 309-
         320 in Finite Simple Groups II (ed. M.J. Collins). NY:
         Academic Press.
[Tits4] Tits, J. (1984). Théorie des groupes (Lectures 1983-84), Annu-
         aire du College de France.
[V]   van Maldeghem, H. (1984). Valuations on PTRs induced by triangle
         buildings. Preprint, Vrije Univ. Gent.
[Wa]   Walker, J. (1982). A poset which is shellable but not lexico-
         graphically shellable. Preprint, Caltech.
[We]   Webb, P. (1983). A local method in group cohomology. Preprint,
         U. Manchester, 1983. Complexes, group cohomology, and an in-
         duction theorem for the Green ring. Preprint, U. Manchester,
         1985.
[Ws]   Wester, M. (1985). Endliche fahnentransitive Tits-Geometrien und
         ihre universellen Uberlagerungen. Mitt. Math. Sem. Giessen,
         170, 1-43.

PROBLEMS PRESENTED AT THE CONFERENCE

N. Alon

Tel Aviv university & Bell communications research

Let $\mu_1,\mu_2,\ldots,\mu_t$ be t continuous probability measures on $I = [0,1]$. An α-*share* is a subset A of I such that $\mu_i(A) = \alpha \;\forall\; 1 \leq i \leq t$. If A is a union of a finite number s of pairwise disjoint intervals, then the *size* of A is s. Otherwise, the size is ∞. For $t \geq 1$ and $0 \leq \alpha \leq 1$ let $f = f(t,\alpha)$ be the minimum (possibly ∞) f such that for every $\mu_1,\ldots,\mu_t$ as above there is an α-share of size at most f.

<u>Problem</u>: Is $f(t,\alpha) < \infty \quad \forall\; t,\alpha$?

(Known for all (t,α) with $\alpha = $ rational and for all (t,α) with $t \leq 2$.)

P.J. Cameron
Oxford

a) Let S be a subset of $\mathbf{Z}/(n)$. We say that S is

sumfree if $x,y \in S \Rightarrow x + y \notin S$;

complete if $z \notin S \Rightarrow \exists\, x,y \in S$ with $x + y = z$.

An element $x \in S$ is called *odd* if $-x \notin S$.

i) Show that, for all $n \geq n_0$, there is a complete sumfree set modulo n containing an odd element.

ii) If S^- is the set of odd elements in S, find the supremum of $|S^-| / |S|$ over all complete sumfree sets of residues.

b) Let p be an odd prime. Let V denote the vector space over $GF(2)$ consisting of all binary words of length p. Let $f(p)$ denote the largest codimension of a subspace W of V such that the union of all cyclic shifts of W is V.

Does $f(p)$ tend to ∞ as $p \to \infty$?

(It is known that $f(p) \geq 2$ for all $p \geq 5$. However, only finitely many primes p are known for which $f(p) > 2$ - these include all Fermat and Mersenne primes greater than 7.)

M. Deza
CNRS, Paris, and University of Tokyo

1) Does there exist a geometric lattice of rank 4 with all lines of size 3 and all planes of size 13? The first candidate has $183 = 13^2 + 13 + 1$ points.

2) Let G be a subgroup of S_n^d and

$$L = \{ |\bigcap_{i=1}^{d} F(g_i)| : g = (g_1,...,g_d) \in S_n^d - <1> \} .$$

<u>Conjecture:</u> $|G| \leq \Pi_{\ell \in L} (n-\ell)^d$.

For $d = 1$ it was proved by Kyota.

3) We say that a metric space (X,d) is $(2n+1)$-gonal if

$$\sum_{i,j \in N} d_{ij} + \sum_{i,j \in N'} d_{ij} \leq \sum_{i \in N, j \in N'} d_{ij}$$

(here d_{ij} denotes $d(x_i,x_j)$) for all n points $N \subset X$ and all $n + 1$ points $N' \subset X$ (not necessarily distinct). I proved that (X,d) with $|X| = 5$ embeds isometrically into L^1 iff it is 5-gonal; for all $n \geq 2$ D. Avis gave an example of $(2n+1)$-gonal metric space $(X,d$ with $|X| = 7$, which does not isometrically embed into L^1.

<u>Conjecture</u>: Any (X,d) with $|X| = 6$ embeds isometrically into L^1 iff it is 5- and 7-gonal.

P. Frankl
CNRS Paris

Let p and q be distinct primes and consider the cyclotomic polynomial $h(x) = (x^{pq}-1)(x-1)/(x^p-1)(x^q-1)$.

Suppose that $f(x) = \Sigma_{0 \le i < pq} a_i x^i$ with each a_i being 0 or 1 . Moreover, $h(x) \mid f(x)$.

Does this imply that either $(x^{pq}-1)/(x^p-1)$ or $(x^{pq}-1)/(x^q-1)$ divides $f(x)$?

With R.L. Graham we could prove that the answer is "yes" for $p = 2$ and for $pq = 15,21$.

Since the conference Peter Cameron has provided an affirmative answer.

A.V. Ivanov
Moscow

<u>Conjecture</u>: For distance-regular graph [1] from the conditions

$$(1,a_1,b_1) = (c_{r-1},a_{r-1},b_{r-1}) \ne (c_r,a_r,b_r)$$
$$(c_s,a_s,b_s) = (c_{s+t},d_{s+t},b_{s+t})$$

the inequality $t \le r$ holds

<u>Proved cases</u>:

a) $c_r \geq 2$ P. Terwilliger [2] see also [4]

b) $a_s = 0$ A. Gardiner

c) $c_s = 1$ or $b_s = 1$ A.V. Ivanov [3]

            $c_s > k/2$ or $b_s > k/2$ A.V. Ivanov [4]

            $(k = 1+a_1+b_1)$

<u>References</u>.

1. N.L. Biggs, Automorphic graphs and the Krein condition, Geom. Dedicata 5(1976), 117-27.

2. P. Terwilliger, Distance-regular graphs and (s,c,a,k)-graphs, J. Combin Theory Ser. B 34(1983), 151-86.

3. A.A. Ivanov, A.V. Ivanov and I.A. Faradjev, Distance-transitive graphs of valency 5,6 and 7, European J. Combinatorics 7 (1986), 303-19.

4. A.A. Ivanov, A.V. Ivanov, Distance-transitive graphs of valency k , $8 \leq k \leq 13$, (To appear).

Gil Kalai
Hebrew University

a) A collection **A** of subsets of [n] $(= \{1,2,\ldots,n\})$ is a *refined family* if $\phi \in$ **A** and whenever $S,T \in$ **A** and $S \subset T$ then there exists a chain $S = S_o \subset S_1 \subset \ldots \subset S_\ell = T$ in **A** such that $|S_i| = |S_{i-1}| + 1$, $i = 1,\ldots,\ell$.

Examples:

1) Simplicial complexes C on [n] $(S \in C , R \subset S \Rightarrow R \in C)$.

2) A chain in [n] of size n+1 .

Let f(n) be the number of refined families on [n] . Problem: Estimate $\log f(n)$.

Clearly

$$\log f(n) \ge \binom{n}{[\frac{n}{2}]} \sim C \frac{2^n}{\sqrt{n}}$$

(Kleitman proved that the number of simplicial complexes on [n] is

$$2^{\binom{n}{[\frac{n}{2}]}} (1+o(1)) .$$

I don't know whether $\log f(n) = o(2^n)$.

b) Let L be the family of ranked atomic finite lattices with the property that every interval of rank 2 has exactly 4 elements.

$L \in L$ has property D if there are two co-atoms in L which are "disjoint" (= no atom is covered by both). Prove or disprove: if rank $L \ge 4$ and for every co-atom M of L the lattice [o,M] has property D then L has property D .

A positive answer would imply a conjecture of Kupitz. (The interesting case is for face lattices of polytopes).

c) Let $\ell_1,\ldots,\ell_s$ and $m_1,\ldots,m_s$ be lines in a projective 3-space such that $m_i \cap \ell_j = \phi$ iff $i = j$. Prove or disprove: $s \le 6$. This is true over a field but nothing is known even over the Quaterions. (Over a commutative field it is true (Lovasz 77') that if the ℓ_i's and m_i's are r-dimensional and t-dimensional spaces, resp., of a projective (r+t+1)-space which satisfy the above condition then

$$s \le \binom{r+t+2}{r+1} .$$

This follows easily from exterior algebra argument which applies only to fields.)

Attila Sali
The Ohio State University

Problems

Let X be an n-elements set. F is a set of subsets of X, $|F| \geq n$.

<u>Questions</u>:

1) If k is a constant, independent of n, and $|F^n| \leq k \cdot n$ for $n > n_0$, where $F^n = \{G : \exists\ F_1,...,F_s \in F$ such that $\cap_{i=1}^{s} F_i = G\}$, what should F look like?

2) Prove, that either $|F^n| \geq n \cdot f(n)$ where $f(n) \to \infty$ a suitable function, or $\exists$ $A_1,...,A_s \in F$ with $s > f(n)$, such that $A_i = B_i \cup C_i$ and B_i's are pairwise disjoint, $C_1 \subseteq C_2 \subseteq ... \subseteq C_s$ and $B_i \cap C_j = \phi$.

3) Prove a statement similar to 2, but with another "simple configuration.

J.D.H. Smith
Iowa S. University

(For the background on quasigroup character tables, see K.W. JOHNSON and J.D.H. SMITH, Characters of finite quasigroups, Eur. J. Comb. $\underline{5}$ (1984), 43-50 or J.D.H. SMITH, "Representation Theory of Infinite Groups and Finite Quasigroups", Séminaire de Mathématiques Supérieures, Université de Montréal 1986.)

<u>Problem 1</u>: Recognize which quasigroup character tables are not character tables of loops.

<u>Problem 2</u>: Recognize which quasigroup character tables are not character tables of groups.

Examples:

There are just three character tables of quasigroups of order 5, namely the group character table, the table

$$
\begin{array}{cc}
1 & 1 \\
2 & -1/2
\end{array}
$$

which is the character table of a loop of order 5 (but not of a group of order 5 , of course), and

1	1	1
$\sqrt{2}$	$\sqrt{2}\cos 72^{\circ}$	$-\sqrt{2}\cos 36^{\circ}$
$\sqrt{2}$	$-\sqrt{2}\cos 36^{\circ}$	$\sqrt{2}\cos 72^{\circ}$

which is the character table of a quasigroup of order 5 , but not of any loop.

Paul Terwilliger
Wisconsin

Question 1. Let X be the set of d-dimensional subspaces of the vector space of dimension n over $GF(q)$. X is a metric space if we define the distance between elements x and y of X to be $d\text{-dim}(x \cap y)$. Find an elgant way to characterize those finite metric spaces embeddable in X for some n,d,q .

Question 2. Let Y be a finite set of points on the unit sphere in some Euclidean space $E,<\ >$. For any real numbers r,s such that $-1 \le r,s \le 1$ and for any $x,y \in Y$ let $P_{rs}(x,y) = \Sigma\, z$,the sum being over all $z \in Y$ with $<z,x> = r, <z,y> = s$. Assume $P_{rs}(x,y)$ is linearly dependent on x,y for all r,s and all x,y .

Find all sets Y satisfying the above condition. Examples are the root systems D_4 and E_8 , and the set of minimal length vectors in the Leech lattice.

Shen Hao
Shanghai Jiao Tong University, Shanghai, China

Let p be a prime, $p \geq 5$ and let

$$
A_o = \begin{bmatrix}
a_{11} & a_{12} & \cdots & a_{1,p-1} \\
a_{21} & a_{22} & \cdots & a_{2,p-1} \\
\multicolumn{4}{c}{\cdots\cdots} \\
a_{p-1,1} & a_{p-1,2} & \cdots & a_{p-1,p-1}
\end{bmatrix}
$$

be a Latin square of order $p-1$ on $GF(p) \setminus \{0\}$. Corresponding to each a_{ij} , we define a difference pair $(j-i , a_{ij}-j)$ where $i,j,a_{ij} \in GF(p) \setminus \{0\}$.

Conjecture. For each prime $p \geq 5$ there exists a Latin square A_o of order $p-1$ on $GF(p) \setminus \{0\}$ such that all the $(p-1)^2$ difference pairs are different from each other and different from $(0,0)$. It is true for $p = 5,7,11,19$.

For EU product safety concerns, contact us at Calle de José Abascal, 56–1°,
28003 Madrid, Spain or eugpsr@cambridge.org.

www.ingramcontent.com/pod-product-compliance
Lightning Source LLC
Chambersburg PA
CBHW022123050726
47590CB00002B/377